Dermatokinetics of Therapeutic Agents

Dermatokinetics of Therapeutic Agents

Edited by
S. Narasimha Murthy

CRC Press
Taylor & Francis Group
Boca Raton London New York

CRC Press is an imprint of the
Taylor & Francis Group, an **informa** business

CRC Press
Taylor & Francis Group
6000 Broken Sound Parkway NW, Suite 300
Boca Raton, FL 33487-2742

First issued in paperback 2017

ISBN 13: 978-1-138-11458-6 (pbk)
ISBN 13: 978-1-4398-0477-3 (hbk)

Library of Congress Cataloging-in-Publication Data

Dermatokinetics of therapeutic agents / editor, S. Narasimha Murthy.
    p. ; cm.
  Includes bibliographical references and index.
  ISBN 978-1-4398-0477-3 (hardcover : alk. paper)
  1. Skin absorption. 2. Pharmacokinetics. 3. Dermatopharmacology. I. Murthy, S. Narasimha (Sathyanarayana Narasimha), 1971-
  [DNLM: 1. Dermatologic Agents--pharmacokinetics. 2. Administration, Cutaneous. 3. Drug Delivery Systems--methods. 4. Skin Absorption. 5. Skin Diseases--drug therapy. QV 60]

RM303.D47 2011
615'.7--dc23
2011018877

Visit the Taylor & Francis Web site at
http://www.taylorandfrancis.com

and the CRC Press Web site at
http://www.crcpress.com

Dedicated to My Parents

Contents

Contents

Contributor Bios

Yuri G. Anissimov, MSc (physics), PhD (applied mathematics), is a senior lecturer of mathematics at the School of Biomolecular and Physical Sciences, Griffith University, Australia. His main research area is mathematical modeling of drug transport through skin and liver kinetics.

April C. Braddy, PhD, is currently a team leader in the Division of Bioequivalence in the Office of Generic Drugs, Center for Drug Evaluation and Research, U.S. FDA. She joined the Division of Bioequivalence as a reviewer in 2006 and has been serving as a team leader since 2008. Dr. Braddy received her BSc in microbiology from Clemson University in 2000 and her PhD in pharmacy (pharmaceutical sciences) from the University of Florida, in 2004. She has been involved in several working groups, which have been tasked with the review of bioequivalence regulatory submissions for several therapeutic drug classes that include dermatologics. She has also been involved in several research projects with a focus on pharmacodynamic modeling using WinNonlin and NONMEM software. Dr. Braddy has coauthored several peer-reviewed research articles and book chapters, and has also served as a reviewer for several journals in the field of medicinal chemistry.

Dale P. Conner, PharmD, is currently the director of the Division of Bioequivalence 1 Office of Generic Drugs, FDA, Rockville, Maryland. Dr. Conner received his BSc in pharmacy from the Massachusetts College of Pharmacy in 1979 and his PharmD from the University of Florida, in 1983. From 1983 to 1985, he completed a postdoctoral fellowship in clinical pharmacology in the Division of Clinical Pharmacology of Thomas Jefferson University, Pennsylvania. He then joined the faculty of the Uniformed Services University of the Health Sciences as an assistant, and later associate, professor in the Division of Clinical Pharmacology. From 1992 to 1994, he was the director of pharmacokinetics for Scios Nova, Inc. From 1995 to 1997, he held the position at FDA of team leader for clinical pharmacology and biopharmaceutics in the areas of pulmonary, allergy, drug abuse, anesthesia, and critical care drug products. Dr. Conner is board certified in applied pharmacology by the American Board of Clinical Pharmacology. His research interests have included pharmacokinetics, drug metabolism, analytical methods, transcutaneous measurement of drugs, measurement of drug effects on the skin, development of nasal and inhalation products, and drug therapy of sepsis and ARDS.

M. Begoña Delgado-Charro, PhD, qualified as a pharmacist and received her PhD from the University of Santiago de Compostela (1990). She was an MEC-Fulbright fellow at University of California, San Francisco (1991–1993), where she specialized in iontophoresis. She worked at the Universities of Santiago de Compostela and Geneva and moved to the University of Bath, United Kingdom, in 2004, where she currently works as a senior lecturer. Her teaching activity concerns pharmaceutics, drug delivery, and pharmacokinetics. Her research involves the use of iontophoresis to optimize transdermal drug delivery, drug delivery to the nail and as a tool for noninvasive drug monitoring and pharmacokinetics. Her work has been funded by the Swiss National Science Foundation, Parkinson's Disease Society, United Kingdom, MRC, NHS-NIC, and by several pharmaceutical companies. Dr. Delgado-Charro is a member of the IATDMCT, EUFEPS, APV, APS, CRS and the Medicines for Children Research Network. She has published 68 peer-reviewed articles in the scientific literature, 9 book chapters, 100 congress abstracts, and has several patents in the field of iontophoresis.

Hartmut Derendorf, PhD, is a distinguished professor and chairman of the Department of Pharmaceutics at the University of Florida College of Pharmacy. Professor Derendorf has published over 340 scientific publications and 6 textbooks in English and German. He is an associate editor of the *Journal of Clinical Pharmacology*, the *European Journal of Pharmaceutical Sciences*, and the editor of the *International Journal of Clinical Pharmacology and Therapeutics* and *Die Pharmazie*. Professor Derendorf has served as president of the American College of Clinical Pharmacology (ACCP) and president of the International Society of Anti-Infective Pharmacology (ISAP). He was awarded the Distinguished Research Award and the Nathaniel T. Kwit Distinguished Service Award of ACCP, the Research Achievement Award in Clinical Science of the American Association of Pharmaceutical Sciences (AAPS), and the Volwiler Award of the American Association of Colleges of Pharmacy (AACP).

José Juan Escobar-Chávez, PhD, is a professor at the Faculty of Chemistry and Pharmacy, University of Mexico (UNAM). His research interests are in the topical and transdermal delivery of drugs and include the use of tape-stripping technique to quantify the amount of drug in layers of stratum corneum, the use of TEWL and ATR/FTIR to show the changes of skin as a result of the use of chemical enhancers, the use of iontophoresis and microneedles in the transdermal administration of drugs, and the development and characterization of transdermal patches and gels. The application of nanotechnology in drug delivery is another major research interest. He has published a number of research papers in these areas. He has written three book chapters and edited an e-book with Bentham Science Publishers.

Miriam López-Cervantes, PhD, is a professor at the Faculty of Chemistry and Pharmacy, University of Mexico, and qualifier at Comisión Federal de Prevención contra Riesgos Sanitarios (COFEPRIS). Dr. López-Cervantes's research interests are in the topical and transdermal delivery of drugs and include the use of the chemical and physical enhancers for the absorption of substances through the skin, like laurocapram and its derivatives, and development and characterization of transdermal patches and gels.

S. Narasimha Murthy, PhD, is an assistant professor of pharmaceutics at the University of Mississippi. He received his PhD in pharmaceutics from Bangalore University, India, in 2003. He completed his postdoctoral research at Roswell Park Cancer Institute, New York, in 2005. His research interests are mainly in the area transdermal and trans-ungual drug delivery. Dr. Murthy has developed several innovative technologies to enhance drug delivery across the skin and nail plate. He has published over 60 research papers in peer-reviewed scientific journals. He is the editor of the book *Dermatokinetics of Therapeutic Agents.*

Michael A. Repka, PhD, is chair and associate professor of the Department of Pharmaceutics at the University of Mississippi, as well as director, Center for Thermal Pharmaceutical Processing. He joined the faculty at University of Mississippi after receiving his PhD from the University of Texas College of Pharmacy. His research interests include oral transmucosal and transdermal/trans-nail delivery systems. Many of these systems are directed toward the solubilization and delivery of poorly soluble bioactives via hot-melt extrusion technology, which is a primary focus of his research. Polymeric drug delivery design, stabilization of conventional and novel drug delivery systems, formulation and process development for natural products such as tetrahydrocannabinol and pro-drugs, in addition to antifungal/antibacterial agents and vaccines, are also a significant part of his research. Dr. Repka's publications include over 60 peer-reviewed journal articles and well over 200 presentations at national/international scientific meetings.

Michael Roberts, B Pharm, MSc, PhD, DSc, MBA, FACP, is an Australian National Health and Medical Research senior principal research fellow, professor of therapeutics and pharmaceutical science at the University of South Australia (UniSA), and a professor of clinical pharmacology and therapeutics in the School of Medicine at the University of Queensland (UQ). He is director of the Therapeutics Research Centre with the UniSA Unit based at the Queen Elizabeth Hospital in Adelaide and the UQ Unit based at the Princess Alexandra Hospital (PAH) in Brisbane. He has more than 350 peer-reviewed research publications (more than 625 communications in total, including 46 book chapters) and is the coeditor of three research books on topical drug delivery as well as three others. Two of his main research interests are topical drug delivery and pharmacokinetics.

Flora Adriana Ganem Rondero, PhD, obtained his PhD in pharmaceutical sciences at the University of Geneva (Switzerland) and Claude Bernard–Lyon I (France). Since 1988, Dr. Rondero has worked in pharmaceutical technology at Universidad Nacional Autónoma de México. She was appointed in the current position in the laboratory of pharmaceutical technology (Facultad de Estudios Superiores Cuautitlán, Universidad Nacional Autónoma de México) in 1998. Her research interests are in the area of permeation enhancers in topical and transdermal drug delivery.

Srinivasa Murthy Sammeta is a PhD student of Dr. S. Narasimha Murthy in the Department of Pharmaceutics at the University of Mississippi. Sammeta has published 15 research papers on topics related to transdermal and topical drug delivery and sampling. He is an NIH pre-doctoral fellow, and a member of the American Association of Pharmaceutical Scientists and Rho Chi. Sammeta received his master's degree in 2006 from Rajiv Gandhi University of Health Sciences in India.

Rong Shi received her master of science degree in chemistry from the University of Missouri-Rolla (now Missouri University of Science and Technology). Since 2006, she has been pursuing her doctoral research in pharmaceutics at the University of Florida, under the supervision of Dr. Hartmut Derendorf.

Grazia Stagni, MS, PhD, is an associate professor of pharmaceutics at Long Island University, Brooklyn, New York. She received her laurea in medicinal chemistry and pharmaceutical technology from the Università degli Studi of Bologna in Italy and her master's and PhD in pharmaceutics from the University of Texas at Austin. After graduation, she joined the Division of Clinical Pharmacology at the University of Texas Health Science Research Center in San Antonio as a research fellow. In 2000, Dr. Stagni joined the faculty at Long Island University. Her research interests include pharmacokinetics and pharmacodynamics, microdialysis in skin, and dermal and transdermal drug delivery.

Georgios N. Stamatas, PhD, graduated with honors from the Department of Chemical Engineering at the Aristotle University of Thessaloniki, Greece. He received his PhD in chemical/biomedical engineering from Rice University, Houston, Texas. Following a postdoctoral fellowship in tissue engineering at Rice University, Dr. Stamatas joined the R&D group of Johnson & Johnson Consumer Products at Skillman, New Jersey, and later at Issy-les-Moulineaux, France. His research focuses on biomedical applications in dermatology with an emphasis on skin biophysics for the development of noninvasive in vivo methods. His current interests include skin physiology and development of infant skin during the first years of life. Dr. Stamatas has authored over 40 scientific publications, over 100 meeting abstracts, and 8 international patent applications.

Ya-Ting Wu is a PhD student at the University of Queensland, Australia. Her research interests include topics related to drug delivery to the hair follicles and in developing a novel treatment for common skin diseases such as hair loss, acne, and rosacea.

1 Introduction to Dermatokinetics

Ya-Ting Wu, Yuri G. Anissimov,
and Michael S. Roberts

CONTENTS

1.1 INTRODUCTION

The skin is one of our key defensive barriers that protects the internal living organism from the external environment and assists in maintaining its homeostasis. Although a key role of its outermost layer is to prevent the entry of foreign substances into the body, this barrier is imperfect and can allow certain topically applied substances. In addition, substances entering the body via the skin avoid first-pass hepatic metabolism

1

and gastric irritation. The skin also allows easy access for a topically applied product with the advantage of easy removal if side effects are to occur. For example unwanted dreams with nicotine patches necessitate patch removal. The skin is, therefore, regarded as an important route for systemic drug delivery, and, at the same time, a key barrier to the ingress of hazardous materials from our external environment.

Dermatokinetics, the kinetic processes associated with skin absorption, is frequently used to determine the efficacy and toxicity of topically applied substances. Any study on the dermatokinetics for a particular substance is likely to include the following [1]:

1. Skin exposure to a substance in a particular form, that is, in cream, formulations, patches
2. Entry through the skin by the stratum corneum or appendages (the route of penetration)
3. Transport through various skin layers into the blood or lymphatics
4. Entry into the systemic circulation where distributed elsewhere in body, metabolized, and/or excreted (pharmacokinetics)
5. Interaction of the penetrating substance with the cells in the skin and/or in the body

In this chapter, we introduce the concept of dermatokinetics. We begin with the skin structure and its effect on the penetration of topically applied substances. Figure 1.1 is an overview of various processes when skin is exposed to different molecules.

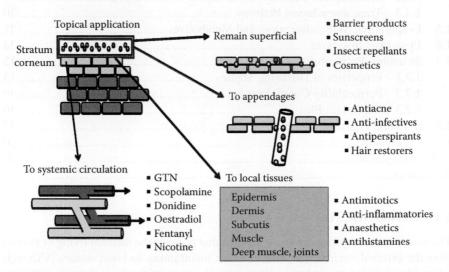

FIGURE 1.1   (**See color insert.**) Schematic diagram showing various processes when the skin is exposed to different molecules. (Modified from Roberts, M.S. et al., Skin transport, In: Walters, K.A., editor, *Dermatological and Transdermal Formulations*, Marcel Dekker, New York, 2002, pp. 89–195. With permission.)

1.2 SKIN STRUCTURE

1.2.1 FUNCTION OF THE SKIN

The skin is the body's largest organ, accounting for more than 10% of body mass. Figure 1.2 shows the histology of thick (palm) and thin (scalp) skin. It is evident that the main appendages for the palm are the sweat ducts, whereas those for the scalp are the hair follicles. The skin consists of four layers: the stratum corneum (nonviable epidermis, abbreviated as SC); the remaining layers of the epidermis (viable epidermis); dermis; and subcutaneous tissues. The appendages are the hair follicles, sweat ducts, apocrine glands, and, on the hands and feet, nails (hardened keratin). The skin protects the body against infection and solar radiation, maintains homeostasis (heat regulation), and, because of its rich enervation, senses the environment. As shown in Figure 1.2, the skin may vary greatly between sites and is thinner on the scalp and face but thicker on friction-bearing regions like the palms of the hands and soles of the feet.

1.2.2 STRATUM CORNEUM

The outer layer of the epidermis is the SC, also known as the horny layer or keratin layer. SC is a nonviable epidermis, consisting in the cross section of 15–25 flattened, stacked, hexagonal, and cornified cells. Each cell is about 40 µm in diameter and 0.5 µm in thickness and composed of insoluble bundled keratins (about 70%) and lipids (about 20%) encased in a cell envelope, accounting for about 5% of the SC weight. The SC represents the end phase of skin differentiation and is essentially a lipid–protein biphasic structure [2,3]; for most solutes, it is the main barrier to penetration. Figure 1.3 is a diagrammatic representation of the structure of the SC.

The SC comprises columns of tightly packed corneocytes organized into clusters of up to a dozen cells per corneocyte layer. The laterally overlapping, quasi-columnar stacks of cells in the SC are sealed tightly by densely packed lipid multilayers, and,

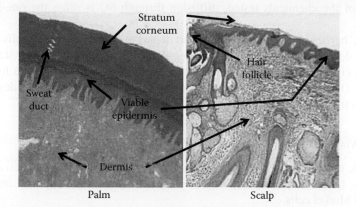

FIGURE 1.2 (See color insert.) Structure of mammalian skin. (With permission of Allan F. Wiechmann, reproduction of slides 44 and supplemental slide 101 at www.ouhsc.edu/histology/, accessed March 2010. With permission.)

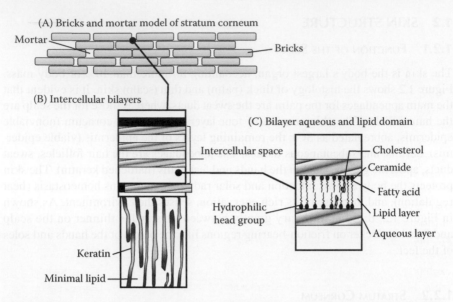

FIGURE 1.3 Diagrammatic representation of the structure of the SC showing (A) the bricks and mortar model of its gross structure, (B) the intercellular bilayers, and (C) the location of polar and lipid domains.

therefore, it is often described as "bricks and mortar" (analogous to a wall) [4]. The corneocytes, containing hydrated keratin, comprise the "bricks," embedded in a "mortar," composed of multiple lipid bilayers of ceramides, fatty acids, cholesterol, and liquid crystals domains [5]. The "mortar-like" lipid bilayers are known to be the important regulators of skin permeability. Specifically, drug permeability is reported to be inversely proportional to the relative lipid content in the SC [6]. In addition, lipids also have a critical role in the regulation of SC cohesion and desquamation [7]. The dead and flattened cells in the SC protect the deeper tissue, and, thus, for the majority of the chemicals tested, diffusion through SC is often the rate-limiting step [8,9].

The SC barrier properties may be related in part to a high turnover of cells through continual desquamation; very high packing density ($1.4\,g/cm^3$ in the dry state); low hydration (15%–10% compared to the usual 70%); and low surface area for solute transport. Disorders of epithelialization such as psoriasis lead to a faster desquamation, reducing the effectiveness of the SC as a barrier.

1.2.3 VIABLE EPIDERMIS

The viable epidermis is typically around $200\,\mu m$ in thickness and lies at a depth of $10–15\,\mu m$. It contains living cells such as keratinocytes, melanocytes, Langerhans cells, and Merkel cells.

Its functions include supply of cells to form the SC; metabolism of substrates (including exogenous substances); synthesis of melanin for skin pigmentation and sun protection; immune response; and sensory perception.

1.2.4 DERMIS

The dermis is five to seven times thicker than the epidermis (1–1.5 mm) and lies below the epidermis that is connected to the dermis by the "basement membrane." It is a tough layer of horizontally arranged collagen and elastic fibers, with fibroblasts [10]. The dermis has numerous structures embedded within it: blood and lymphatic vessels, nerve endings, pilosebaceous units (hair follicles and sebaceous glands), and sweat glands (exocrine and apocrine). This layer is often viewed as essentially gelled water, and thus provides maximal cushioning but a minimal barrier to the delivery of polar and moderately lipophilic drugs, although the dermal barrier may be significant when delivering highly lipophilic molecules [11].

The main structural component of the dermis is referred to as the coarse reticular layer. There is an extensive vascular network in the dermis providing for skin nutrition, repair, and localized immune responses, as well as allowing heat exchange and connecting the skin to the systemic immune system. The blood flow rate to the skin is about 0.05 mL/min/cc of skin [12], providing a vascular exchange area equivalent to that of the skin surface area. Skin blood vessels derive from those in the subcutaneous tissues with an arterial network supplying the papillary layer, the hair follicles, the sweat and apocrine glands, and the subcutaneous area as well as the dermis itself. These arteries feed into arterioles, capillaries, venules, and finally veins. Of particular importance in this vascular network is the presence of arteriovenous anastomoses at all levels in the skin. These arteriovenous anastomoses, which allow a direct shunting of up to 60% of the skin blood flow between the arteries and veins and thus avoid the fine capillary network, are critical to the skin's functions of heat regulation and blood vessel control. Blood flow changes are most evident in the skin in relation to psychological effects such as shock (draining of color from the skin) and embarrassment (blushing), temperature effects, and physiological responses to exercise, hemorrhage, and alcohol consumption.

1.3 MATHEMATICAL DESCRIPTION OF DRUG TRANSPORT THROUGH BARRIERS

As discussed in Section 1.2, skin is a complex multilayered structure that can be modeled as a stack of homogeneous barriers. In this section, we will consider simple modeling of transport through such barriers.

1.3.1 SOLUTE TRANSPORT THROUGH ONE BARRIER

Figure 1.4A shows an initial application of the solute to the barrier. In this case, solute molecules are concentrated at the top of, but just inside, the membrane. After a long time has passed (steady state), providing the solute molecules are removed at the bottom of the barrier, the linear concentration profile will form in the membrane (Figure 1.4B). The rate of transport through the membrane at steady state (called flux, J_{ss}, has dimension of mole/cm²/s or g/cm²/s) is determined by the solute diffusion coefficient (D, has the dimension of cm²/s), the

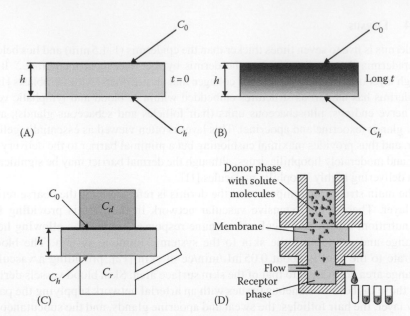

FIGURE 1.4 Transport is determined by a chemical potential in a membrane; at $t = 0$, all of the potential (concentration) is at the top surface and nil elsewhere. (A) After some time, the chemical will diffuse to reach equilibrium across the membrane thickness h, (B) in this case, the highest concentration C_0 is at the source at the top and the lowest C_h at the bottom at the sink. It is measured by applying a donor concentration C_d and measuring a receptor concentration C_r in either (C) static "Franz" diffusion cell or (D) flow-through diffusion cells.

difference between concentrations at the top (C_0) and bottom (C_h) of the barrier and the thickness of the barrier (Fick's law) [13]:

$$J_{ss} = \frac{D(C_0 - C_h)}{h} \tag{1.1}$$

Figure 1.4C is the schematic diagram of vehicle/donor and receptor application to the membrane for a static Franz cell experiment and Figure 1.4D is the schematic of the flow through diffusion cell penetration experiment. The receptor provides removal of the solute molecules at the bottom of the membrane. The solute now has to partition from the vehicle/donor into the membrane to start diffusion in the membrane before removal into the receptor. This process is defined by the partition coefficient (K) that is the ratio between the membrane concentration (C_m^{ss}) and the donor phase concentration (C_d^{ss}) at the steady state: $K = C_m^{ss}/C_d^{ss}$. Provided that the donor phase is not significantly depleted during the transport process (this situation is referred to as an infinite donor), the concentration in the donor will remain constant and the concentration at the top of the barrier will be $C_0 = KC_d$. Similarly, the concentration at the bottom of the membrane will be $C_h = K_r C_r$, where K_r is the partition coefficient between the membrane and the receptor phase. The equation for the steady-state flux then becomes

$$J_{ss} = k_p\left(C_d - \frac{K_r}{K}C_r\right) \tag{1.2}$$

where k_p is the permeability coefficient of the barrier:

$$k_p = \frac{KD}{h} \tag{1.3}$$

To provide efficient removal of the solute molecules, the receptor solution concentration at $t = 0$ must be zero. In many cases, the receptor compartment is large enough or is renewed during the experiment, so that the receptor solution concentration stays close to zero (or more precisely $K_r/KC_r \ll C_d$, which is referred to as sink condition), so that Equation 1.2 becomes

$$J_{ss} = k_p C_d \tag{1.4}$$

Using this equation, the amount of solute penetrating through the membrane would be

$$Q(t) = J_{ss}At = k_p C_d At \tag{1.5}$$

where A is the area of application. Equation 1.5 is a reasonable approximation at very long times, but it does not take into account the fact that at early stages of the experiment the flux is significantly less than the steady-state flux defined by Equation 1.4. When the transient stage is taken into account, the amount of a solute absorbed at time t into the receptor phase is defined by [14]

$$Q(t) = J_{ss}A\left(t - lag - \frac{2h^2}{\pi^2 D} \sum_{n=1}^{\infty} \frac{(-1)^n}{n^2} \exp\left(-t\frac{D}{h^2}\pi^2 n^2\right) \right) \tag{1.6}$$

where lag is the lag time given by

$$lag = \frac{h^2}{6D} \tag{1.7}$$

Figure 1.5 shows the graph of $Q(t)$ as defined by Equation 1.6. It can be noted that at long times (practically for $t > 2lag$), the curve is approaching the line (shown as a dashed line). The explanation to this limiting behavior is that the summation term in Equation 1.6 can be ignored at long times (as the exponent

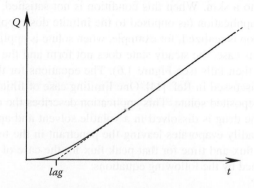

FIGURE 1.5 The amount of a solute absorbed at time t versus time (solid line, Equation 1.6) and the steady-state approximation (dashed line, Equation 1.8).

of a very large negative number quickly approaches zero), so that Equation 1.6 reduces to the form of Equation 1.8:

$$Q(t) = J_{ss}A(t - lag) = k_p C_d A(t - lag) \tag{1.8}$$

Equation 1.8 represents a line (dashed line in Figure 1.5), which for $t > 2lag$ gives a good approximation to the amount of solute penetrating though the membrane.

1.3.2 SOLUTE TRANSPORT THROUGH MULTIPLE BARRIERS

In reality, the skin is not a single barrier but a composition of barriers such as SC, viable epidermis, and dermis (see Figure 1.7). In this case, the total permeability of SC, viable epidermis, and dermis will be determined by

$$\frac{1}{k_p^{total}} = \frac{1}{k_p^{sc}} + \frac{1}{k_p^{ve}} + \frac{1}{k_p^{d}} \tag{1.9}$$

where k_p^{sc}, k_p^{ve}, and k_p^{d} are permeability coefficients of SC, viable epidermis, and dermis, respectively. Note that Equation 1.9 is analogous to determining the electrical conductivity of three resistors in series (conductivity of a resistor is the reciprocal of its resistivity).

When two or more transport pathways exist through SC (as shown in Figure 1.7), the combined permeability is the sum of permeability coefficients:

$$k_p^{sc} = k_p^{lipid} + k_p^{trans} + k_p^{app} \tag{1.10}$$

where k_p^{lipid}, k_p^{trans}, and k_p^{app} are the permeability coefficients through lipid, transcellular, and appendageal pathways through the SC. Note that Equation 1.10 is analogous to determining the electric conductivity of three resistors in parallel. The above pathways are, therefore, referred to as parallel pathways through the SC.

1.3.3 FINITE DOSE DONOR/VEHICLE

The total amount absorbed is described by Equation 1.6 only if depletion in the donor phase is negligible, that is, the amount absorbed is much less than the total dose of the solute applied to a skin. When this condition is not satisfied, it is said that we have a finite dose application (as opposed to the infinite dose application discussed above). This situation is realized, for example, when solute is applied as an ointment at $10\,\mu L/cm^2$. In this case, the steady state does not form and the flux of the solute first increases and then falls (see Figure 1.6). The equations for this case are more complex and are discussed in Ref. [14]. One limiting case of finite dose application is that of solvent-deposited solute. This application describes the condition where a small amount of the drug is dissolved in a volatile solvent and applied to the skin; the solvent then readily evaporates leaving the penetrant in the top layer of the SC surface. The peak flux and time for that peak flux for the case of solvent-deposited solute are determined by the following equations:

$$J_p = \frac{1.85 doseD}{h^2}, \quad t_p = \frac{h^2}{6D} = lag \tag{1.11}$$

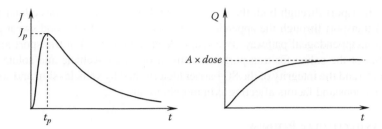

FIGURE 1.6 **(See color insert.)** The flux of solute and the amount of a solute absorbed versus time for finite vehicle application.

where

J_p is the peak flux

t_p is the time of the peak flux

dose is the dose of the solute deposited per area of the skin

The peak flux (J_p) of a solute is important in determining the maximal acute dermal toxic or systemic effects. In practice equations 11 are only simplifications as solute slow binding and incomplete dose availability will affect the peak flux and the time of the peak flux. As shown in Figure 1.6, the total amount absorbed ($A \times dose$ expressed per unit area) defines the total exposure.

1.4   SKIN MORPHOLOGY AND TRANSPORT PATHWAYS

Figure 1.7 is a schematic diagram showing the pathways by which a solute may traverse from the skin surface to the underlying layers and finally into the systemic circulation. There are three pathways: through the lipids in the SC—often referred to as intercellular

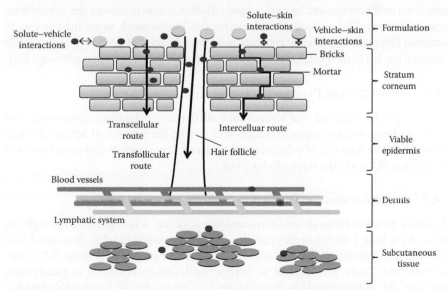

FIGURE 1.7   Schematic diagram showing three principal penetration pathways.

pathway; transport through both the lipids and the keratinocytes—transcellular pathway; and transport through the appendages such as the hair follicles and sweat gland ducts—transappendageal pathway. Two major determinants in skin transport are the availability of the solute for transport (determined by solute–vehicle and solute–skin interactions) and the integrity of the SC barrier (determined by vehicle–skin and solute–skin interactions and factors affecting skin morphology).

1.4.1 INTERCELLULAR PATHWAY

There is continual desquamation of the SC, with a total cellular turnover occurring once every 2–3 weeks in normal humans and faster in certain skin diseases such as psoriasis. Given the high turnover, lipophilic agents such as sunscreens and substances bound to SC (e.g., hexachlorophane) may be less effectively absorbed into the body than would be indicated by the initial partitioning of the agents into the SC layer from an applied vehicle. Indeed, it is now recognized that most solutes enter the body via intercellular regions of the SC that are <0.1 μm wide [1]. The intercellular route is shown in Figure 1.7, where molecules travel in the "mortar," which are the lipoidal domains of the SC. Passive percutaneous transport, therefore, depends on solute properties including the drug's molecular size, lipophilicity, charge, and melting point and how they interact with the skin under various conditions and after application of various formulations.

The intercellular SC spaces were initially not recognized as a significant diffusion pathway because of the small volume they occupy [13]. However, tracer radiolabeled studies [15,16] have shown that solutes are mainly transported in the lipid in these intercellular SC spaces, and not via the corneocytes [17]. Histochemical studies have also shown that the intracellular spaces of the SC squamae or "bricks" are devoid of lipid [18,19], and, that, since the majority of lipids present in other regions is highly nonpolar, there is no structure suitable to form a lipid diffusional matrix around the intracellular keratin filaments. The intercellular volume fraction is also much larger than originally estimated [20] and experimental evidence using precipitation of percutaneously applied n-butanol has led to the visualization of permeation through intercellular pathways [21].

1.4.2 TRANSCELLULAR PATHWAY

It was originally believed that transcellular diffusion mechanisms dominated over the intercellular and transappendageal routes during the passage of solutes through the SC [22]. The majority of experimental evidence now suggests that most transport through the SC is via the intercellular route.

1.4.3 TRANSAPPENDAGEAL PATHWAY

The third possible route of solute transport across the skin barrier is through the walls of the hair follicles or through the sebaceous glands [23,24]. Fractional hair follicular area relative to skin surface available for transport is only about 0.1%, and, therefore, this route was thought to have a negligible contribution to steady-state drug flux [5]. Scheuplein [25], formulating his "principles of transient diffusion," suggested that in the initial or nonsteady-state phase of diffusion, the influence of the transfollicular route is significant for most small nonelectrolytes. Specifically, Scheuplein [25,26] indicated that during the initial phase of diffusion, only transient diffusion occurs, but after steady-state diffusion is established, bulk diffusion through the lipid–keratin matrix of the SC becomes dominant.

Based on Scheuplein's theory, investigators have improved and developed quantitative methods for assessing follicular delivery. Grams and coworkers [27] studied diffusion of a lipophilic dye (Bodypy® FL C5) into the hair follicles of human skin in vitro using real-time online confocal microscopy and observed a high variation in the follicular and the non-follicular regions within the initial phase of diffusion (10 min) with the variation strongly reduced after 40 min. This suggests that for the first 10 min during the transient diffusion phase the follicular route is dominant, while bulk diffusion via other routes becomes fully established after 40 min (reaching steady-state diffusion).

Grams et al. [28] investigated the influence of permeant lipophilicity and vehicle composition on local accumulation using diffusion studies and confocal laser scanning microscopy. They found that formulations saturated with the poorly lipophilic Bodypy dye have a higher flux and higher follicular accumulation in citric acid buffer (compared to 8% surfactants and 20% propylene glycol). Highly lipophilic Bodypy dyes tested had a higher flux than low lipophilicity dyes; and the flux and follicular accumulation were increased after the addition of surfactants (sodium laureth sulfate and cocamidopropyl betaine) and propylene glycol.

In addition, in a human skin study after follicular casting, Vogt et al. [29] compared particles in the size range of 40 and 750 nm and found that 40 nm particles penetrated deeply into the follicular duct of vellus hair, while 750 nm particles remained in the superficial parts of the infundibulum. It was also found in the in vivo pig skin study by Lademann et al. [30] that nanoparticles predominantly penetrate via hair follicles compared to non-particulate formulations. The penetration of micro- and/nanoparticles through the skin can be enhanced by massage. Toll et al. [31] pretreated the skin with massage to achieve deep penetration and observed that the 320 nm particles may be entrapped underneath the cuticular cells and guided further down along the hair follicle duct as the hair moves back and forth. This concept is later described as the "geared pump hypothesis" by Lademann et al. [32].

1.5 EVAPORATION, SKIN ADSORPTION, AND METABOLISM

Figure 1.8 shows examples of processes that can reduce penetration of solutes through the skin. Some ionic compounds such as surfactants [33] and antiseptics [34] bind very strongly to the SC and are not available for penetration. We, therefore, use an availability term F_a, which is the fraction of drug that is effectively available for penetration through the skin. In addition, as discussed below, metabolism of solutes in the SC and epidermis are clearance mechanisms, which can affect skin permeabilities and resultant pharmacological effects.

The effective availability in the presence of skin metabolism is defined as F_m. Accordingly, if both adsorption/evaporation and metabolism are present in vivo but not in vitro, the in vivo flux can be related to the in vitro skin flux based on loss from the donor phase by $J_{in\,vivo} = F_a F_m J_{donor}$. A second important consideration is the differences between in vivo and in vitro skin fluxes. If there is no residual in vitro skin metabolism,

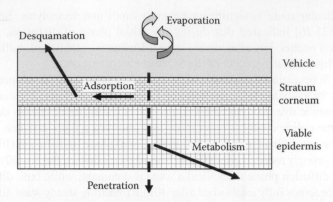

FIGURE 1.8 Schematic diagram showing evaporation, skin adsorption, and metabolism.

then the in vivo flux with metabolism is related to the in vitro flux by $J_{\text{in vivo}} = F_m J_{\text{in vitro}}$. Solute flux through the SC may also be reduced by an inadequate cutaneous blood supply. Other compounds may be lost by evaporation or by desquamation of the SC and are also not available for absorption. These processes will effectively reduce the fraction of the dose applied that is actually absorbed. Hence, the Dose recovered in vivo = $F \times$ Dose applied, where F is the availability and is given by $F = F_a F_m$.

Let us now look at those solutes with significant skin metabolism, that is, $F_m < 1$. The reported metabolic enzyme activity of the skin is about 2% that of the liver. However, the transit time of compounds in the skin is much longer than that in the liver and so the skin may have a much lower F_m than anticipated. Figure 1.9 summarizes the main metabolic enzymes found in the skin and also shows key transporters and the blood and lymph vessels. These metabolic enzymes include the cytochrome P450 CYP1, CYP2, CYP3 enzyme families in the epidermis, hair follicles, and sebaceous glands, as well as flavin-containing monooxygenases, alcohol and aldehyde dehydrogenase enzymes, hydrolase enzymes (glycosidases, phospholipases, sphingomyelinases, phosphatases, esterases, and proteases), and transferases (glutathione-*S*-transferases, *N*-acetyl-transferases, and glucuronosyl transferases) [35]. Also shown in Figure 1.9 are the known transporters expressed in the skin. Enzyme and transporter activities may be modified by other agents applied to the skin. For instance, 0.1% topical retinoic acid cream increases the P450 enzyme activity and human epidermal cells hydroxylate vitamin D3 to calcitriol, the most active form of vitamin D3 and an agent that inhibits cell proliferation but induces differentiation. The cytochrome P450 enzyme, CYP2S1, is induced in human skin by exposure to UV radiation and is upregulated in psoriatic skin and after topical application of retinoic acid. Esterases are especially active in the skin and to the first-pass metabolism of a range of drugs including methyl salicylate, nitroglycerin, and various ester prodrugs [35].

The metabolism of solutes by skin enzymes has also been reviewed by Hotchkiss [36]. In relation to skin transport modeling, difficulties may arise from the as yet undefined anatomical distribution of metabolizing enzymes, both in the various layers of the skin and appendages and the variable activity that may arise from processing skin for permeation studies. For instance, the activity of some enzymes is reduced by the process of heating used to separate the dermis from the epidermis [37]. Some

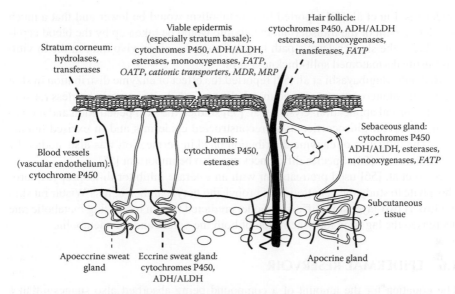

FIGURE 1.9 Main metabolic enzymes found in the skin and their positions. (Adapted from Dancik, Y., Thörling, C., Krishnan, G., and Roberts, M.S., Cutaneous metabolism and active transport in transdermal drug delivery. In *Toxicology of the Skin: Target Organ Toxicology Series*, Monteiro-Riviere, N. (ed.), Taylor & Francis Group LLC, Boca Raton, FL, pp. 69–82, 2010.)

of the key enzymes involved in skin metabolism include aryl hydrocarbon hydroxylase, de-ethylases, hydrolases, monooxygenases, esterases, peptidases, and aminopeptidases. This skin enzyme activity can vary among species and may be induced. A major outcome of these enzymes' activity is the skin first-pass effect whereby a significant proportion of the solute is metabolized between application to the skin and diffusion to its site of action in regions of the skin or into the systemic circulation.

Nakashima et al. [38] used intravenous and transdermal ointment administration of nitroglycerin to estimate the fraction of nitroglycerin avoiding this first pass, which was 0.68–0.76 and comparable to values reported in Rhesus monkeys (0.80–0.84). A higher skin first pass has been reported for methyl salicylate where the first-pass availability in both humans [39] and rats [40] is very low. It has been suggested that the new retinoid, tazarotene, is superior to those used orally because of its limited percutaneous penetration as well as its rapid esterase metabolism in the skin to a more water-soluble active metabolite tazarotenic acid. The latter has a resultant systemic absorption of between 1% (normal) and 5% (psoriasis) on repeated applications [41,42].

Bronaugh et al. [43] have also reviewed some aspects of cutaneous metabolism during in vitro percutaneous absorption. From the perspective of skin transport, skin metabolism can only be adequately modeled using a two-phase model. One of the first studies in this area was that of Ando et al. [44]. Higuchi's group has since reported a number of papers on the effects of skin metabolism on solute transport [45–49]. In a later study, the analysis of the influence of low levels of ethanol on the simultaneous diffusion and metabolism of estradiol with several enzyme distribution models was determined [50]. It was shown that the best model was that for which the enzyme activity resided totally in the epidermis and near the basal layer of the epidermis. Liu et al. [51] reported that metabolism would be lower and that a much smaller amount of the transdermal metabolite would be taken up by the blood capillary due to the shorter dermis path length for permeants in vivo than in the in vitro case using dermatomed split-thickness skin.

Recently, Sugibayashi et al. [52] reported the effect of enzyme distribution in skin on the simultaneous transport and metabolism of ethyl nicotinate in hairless rat skin after its topical application. Gysler et al. [53] studied the skin penetration and metabolism of topical glucocorticoids in reconstructed epidermis and in excised human skin. The influence of enzyme distribution on skin permeation was also studied by Hatanaka et al. [54]. Species differences can also be important [55].

Seko et al. [56] used pretreatment with an esterase inhibitor di-isopropyl fluorophosphate to study the penetration of propyl and butyl paraben across Wistar rat skin in vitro. A two-layer skin diffusion model predicted that an increasing metabolic rate decreased the lag time for penetration of both the parent and the metabolite.

1.6 EPIDERMAL RESERVOIR

The equation for the amount of a compound being absorbed also suggests that a significant concentration of solute at the inner surface of the SC will reduce flux. In most in vitro studies, $C_{sc(inner)}$ is kept minimal by use of appropriate solvents or a large receptor volume to provide sink conditions. However, it should be emphasized that such conditions and those assuming no significant reduction in surface concentration are ideal and will yield higher fluxes than may be observed in reality. Anxiety may, for instance, arise from high reported in vitro penetration fluxes of the lipophlic sunscreens across excised human epidermis using sink conditions when their absorption in vivo may be retarded by a limited solubility in the dermis and blood. Sink conditions may be enhanced in vivo by epidermal metabolism with more than 95% of methyl salicylate being metabolized in human skin in vivo after topical application of a commercial product [57].

It is likely that for most lipophilic solutes an epidermal reservoir will exist after application in vivo for one of three reasons:

- Strong binding in SC.
- There is a substantial time lag for the solute to diffuse through the skin.
- The epidermis and dermis do not act as an efficient sink.

An example of the historical evidence of this effect is illustrated by the work of Vickers in which vasoconstriction was achieved by occluding an area of skin with plastic film 12–14 days after an initial application of steroid under occlusion and a fading of the vasoconstriction on removal of the film at 16 h [58] (Figure 1.10).

Magnusson et al. [59] found that increasing lipophilicity may influence steroid J_{max} through the epidermis and a tendency to decrease penetration rates through the dermis. A potentially important determinant of the reservoir effect is desquamation. However, as shown in Figure 1.11, for a relatively lipophilic compound, corticosterone, desquamation effects are most pronounced when the skin turnover is very high, such as in psoriasis [58].

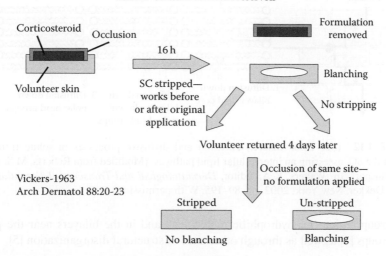

FIGURE 1.10 Skin reservoir effect demonstrated by occlusion of skin previously exposed to corticosteroid cream. (Modified from Roberts, M.S. et al., *Skin Pharmacol. Appl. Skin Physiol.*, 17, 3, 2004. With permission.)

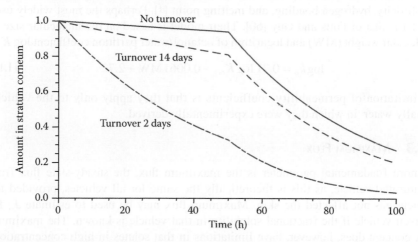

FIGURE 1.11 (**See color insert.**) Amount remaining in SC reservoir of corticosterone with no desquamation (—), a normal epidermal turnover of 14 days (- - -) and a psoriatic epidermal turnover of 2 days (– - –). (Modified from Roberts, M.S. et al., *Skin Pharmacol. Appl. Skin Physiol.*, 17, 3, 2004. With permission.)

1.7 STRUCTURE–PERMEABILITY RELATIONSHIPS

1.7.1 PROPERTIES OF DIFFUSING SOLUTES

Solutes can partition and diffuse through different regions in the lipid bilayers in the SC (see Figure 1.12). As shown in Figure 1.12, solutes can diffuse across SC following a tortuous pathway through the tail-group region (for hydrophobic molecules),

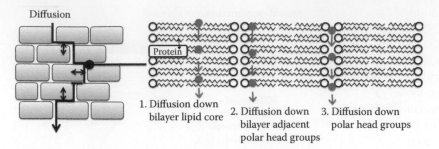

FIGURE 1.12   (See color insert.) Partition and diffusive processes in solute transport through the SC, assessing an intercellular lipid pathway. (Modified from Roberts, M.S. et al., *Skin transport*, In: Walters, K.A., editor, *Dermatological and Transdermal Formulations*, Marcel Dekker, New York, 2002, pp. 89–195. With permission.)

head-group region (for hydrophilic molecules), and in the bilayers near the polar head groups [1] as well as through other sites of structural disorganization [5].

1.7.2   PERMEABILITY COEFFICIENTS

A number of models have been used to relate skin permeability with solute size, liophilicity, hydrogen bonding, and melting point [1]. Perhaps the most widely used model is that of Potts and Guy [60]. Their model is based on the molecular size or molecular weight (MW) and logarithm of octanol/water partition coefficient $\log K_{oct}$:

$$\log k_p = 0.71 \log K_{oct} - 0.0061 \, \text{MW} - 2.72 \tag{1.12}$$

A limitation of permeability coefficients is that they apply only to the vehicle, usually water, in which they were experimentally derived.

1.7.3   MAXIMUM FLUX

A more fundamental parameter is the maximum flux, the steady-state flux from a saturated vehicle, as this is theoretically the same for all vehicles, provided the vehicle has not affected the skin. Maximum flux may be used to estimate J_{ss} for a given vehicle if the fractional solubility in that vehicle is known. The maximum flux concept does, however, have limitations in that solutes in high concentrations may exhibit nonlinear concentration dependencies due to effects on the SC or association in the vehicle. In principle, an ideal maximum flux may be estimated as a product of aqueous solubility and permeability coefficient k_p, where k_p is defined by an expression similar to that developed by Potts and Guy. An alternative approach is to examine experimental or calculated maximum fluxes in terms of their underlying physicochemical determinants. Magnusson et al. [61] investigated the factors that may influence the maximum flux of known drugs and formulations from published literature and found that molecular size (or weight) is the main determinant of solute maximum flux across the skin, while melting point and partition coefficient between octanol and water are less significant. A more recent study by Zhang et al. [62] investigated the relationship between skin solubility and transepidermal flux for similar-size molecules. They showed that a convex dependence of maximum flux on lipophilicity existed and is consistent with a maximum flux occurring between a log K_{oct} of 2 and 3. They showed that the maximum flux was directly related to SC solubility, and not to diffusional or partitioning barrier effects at the SC-viable epidermis interface for the more lipophilic phenols. However, these effects are likely to contribute for very lipophilic solutes. The findings support the previously developed solute structure–skin transport model for aqueous solutions [60] in which it is suggested that skin permeability depends on both partitioning and diffusivity: partitioning is related to K_{oct} and diffusivity to solute size and hydrogen bonding.

Overall, the ideal properties for a molecule to penetrate SC are [5]

- Low molecular mass, preferably less than 600 Da
- Adequate solubility in octanol and water (Log Octanol water—partition coefficient between—1 and 4)
- High but balanced in partition coefficient of solute between membrane and receptor solution
- Low melting point, correlating with good solubility, and minimal number of hydrogen-bonding groups

1.8 SYSTEMIC PHARMACOKINETICS

As discussed earlier, the overall extent of absorption can be defined in terms of the absolute bioavailability, that is, the fraction F of the dose reaching the systemic circulation intact after extravascular administration. This bioavailability is best determined by comparison with plasma/blood or urine levels of unchanged solute achieved with the intravenous administration of the solute. Normally, either the area under the plasma concentration–time profile (AUC) or the amount excreted unchanged in the urine (Au) from the time of administration to infinite time is used to define F:

$$F = \frac{Au_{topical}/Dose_{topical}}{Au_{IV}/Dose_{IV}} = \frac{AUC_{topical}/Dose_{topical}}{AUC_{IV}/Dose_{IV}} \qquad (1.13)$$

Wester et al. estimated F to be 0.57 for nitroglycerin in monkeys using the AUC ratio for unchanged nitroglycerin. A higher F of 0.77 was obtained when the AUC of total radioactivity was used, implying that about 20% of the nitroglycerin had been metabolized during the topical absorption process. This fraction metabolized is a first-pass extraction E_s with $1 - E_s$ defining the skin first-pass availability F_s. The overall availability of intact nitroglycerin F is defined by the product of the fraction released from the product into the SC lipids and undergoing percutaneous absorption FR and F_s, that is, $F = FRF_s$. In this instance, FR is 0.77 and F_s is 0.74. Cross et al. [57] have suggested that in human skin E_s for intact methyl salicylate is >0.95 and hence F_s is <0.05.

One of the earliest detailed evaluations of skin penetration in terms of pharmacokinetic rate constants was presented by Riegelman [63]. Figure 1.13 provides a representation of the events associated with the systemic absorption of a topically applied solute. In the simplest case, absorption is assumed to be determined by the epidermal flux (J_s), the dose applied (*dose*), and F.

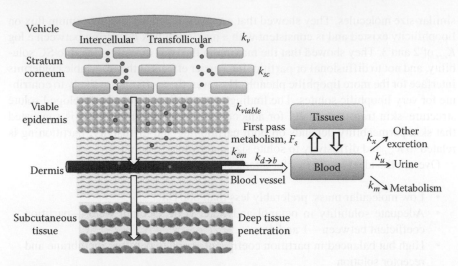

FIGURE 1.13 Schematic representation of the events associated with the systemic absorption of a topically applied solute.

In practice, as described earlier, transport through SC may be associated with a lag time (*lag*), which should be accounted for in pharmacokinetic modeling.

Figure 1.14 shows serum plasma concentrations of testosterone obtained on the application of a patch and on its removal. It is noted that there is a lag prior to reaching maximal levels and in returning to baseline on patch removal. Further, after reaching maximum, the serum levels slowly decline with time, consistent with a reduction in flux due to a gradual depletion in the amount of testosterone

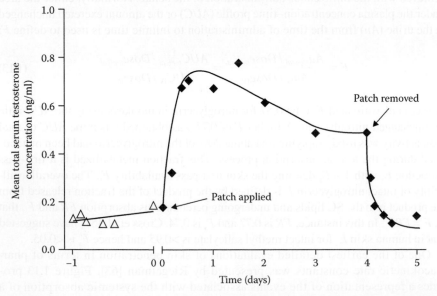

FIGURE 1.14 Mean total testosterone serum concentration over time. (Data from http://www.emea.europa.eu/humandocs/PDFs/EPAR/intrinsa/063406en6.pdf)

TABLE 1.1

Transdermal Delivery—Effective Plasma Concentrations, Systemic Clearances, Estimated F_s and Physicochemical Data Used to Predict Required Solute Transdermal Flux ($J_{s,a}$) (from Equation 1.15)

Solute	Indication	Plasma Level (µg/L)	CL_{body}, L/h/70 kg	Estimated $J_{s,a}$ Required (µg/h)	F_s	$t_{1/2}$ (h)	MW	MP (°C)	log K_{oct}
Buprenorphine	Pain relief	0.1–0.52	76	15–80	0.5	4	468	209	3.44
Clonidine	Hypertension	0.2–2.0	13	2.7–28	1.0	6–20	230	140	1.77
Estradiol	Female hormone replacement	0.04–0.15	4–7 (CL/F)	1–6	0.16	1	272	176	2.69
Ethinyl estradiol	Female contraception	0.011–0.137	28	0.31–3.8	1	17	296	143	4.52
Fentanyl	Chronic pain	1	25–75	25–75	1	17	337	83	4.37
Isosorbide dinitrate	Angina	22	1.22	28	1	105	236	68	1.31
Methyl phenidate	ADHD	20–46	60 L/h/ 30 kg in children	720–2160	1	3–5	233	74.5	2.55
Nicotine	Smoking cessation	10–30	72	900–2630	1	2	162	~80	1.17
Nitroglycerin	Angina	1.2–11	13.5	16.2–148.5	1	0.04	227	13.5	1.62
Norelgestromin	Contraception				1	28	327	131	4.40
Norethindrone acetate	Contraception	2–15	28 (CL/F = 25L/h)	56–420	1	9	340	162	3.99
Oxybutynin	Enuresis	3–4	25–34	75–135	1	2	357	130	5.19
Rivastigmine	Alzheimers	2–10	130	260–1300	1	1–4	250	124	2.14
Rogigotine	Parkinson's disease	0.2–0.6	630	126–378	1	5–7	315	141	4.97
Scopolamine	Motion sickness	0.04	67–205	2.6–8.1	1	1.2–2.9	303	59	1.23
Selegiline	Depression	2			1	2	187	138	2.95
Testosterone	Hypogonadism	60–100	3–5	180–500	1	2.3	288	153	3.31
Timolol	Hypertension	5–15	38	250–750	0.76	4.1	316	72	2.46
Triprolidine	Antihistamine	5–15	43.7	218–655	1	2–6	278	60	4.22

Source: Updated from Roberts, M.S. and Walters, K.A., Human skin morphology and dermal absorption. In: Roberts, M.S. and Walters, K.A., editors, *Dermal Absorption and Toxicity Assessment*, Informa Healthcare, New York, 2009, pp. 1–15.

in the patch. Topical products, especially transdermal patches, often seek to provide constant therapeutically effective plasma concentrations C_{ss}. The in vivo release flux $J_{s,a}$ ideally needed to reach such concentration is given by

$$J_{s,a} = \frac{C_{ss}Cl_{body}}{F_s} \qquad (1.14)$$

where
  Cl_{body} is the total clearance from the body
  F_s is the fraction absorbed into the systemic circulation intact

Table 1.1 [64] shows some estimated transdermal fluxes required for the topical administration of a number of pharmaceuticals that are used in transdermal systems based on the substitution of desired plasma concentrations and body clearances into the equation.

1.9 CONCLUSION

As seen in this chapter, the skin is an important barrier to the ingress of compounds of both therapeutic and potentially toxic compounds. We have attempted to present a review of the current understanding of the factors that affect a solute's ability to traverse the barrier and how these processes can be influenced to enhance penetration. The increasing availability of data from both in vitro and in vivo studies will hopefully make the elucidation of the finer details of these transport processes and the predictability of skin permeation kinetics, through mathematical modeling and interpretation of structure–activity relationships and local physiology, much easier in the years to come. Until such time, however, we hope that the concepts and studies summarized in this chapter will help bring to many an appreciation of the complex nature of skin transport processes and the amount of work that has been involved in bringing us to the level of understanding we have today.

ACKNOWLEDGMENT

This work was supported in part by the National Health & Medical Research Council of Australia.

REFERENCES

1. Roberts MS, Cross SE, Pellett MA. Skin transport. In: Walters KA, editor. *Dermatological and Transdermal Formulations*. New York: Marcel Dekker; 2002. pp. 89–195.
2. Masters BR, So PT. Confocal microscopy and multi-photon excitation microscopy of human skin in vivo. *Optics Express*. 2001;8(1):2–10.
3. Masters BR, Bohnke M. Confocal microscopy of the human cornea in vivo. *Int Ophthalmol*. 2001;23(4–6):199–206.
4. Touitou E. Drug delivery across the skin. *Expert Opin Biol Ther*. 2002;2(7):723–733.
5. Barry BW. Novel mechanisms and devices to enable successful transdermal drug delivery. *Eur J Pharm Sci*. 2001;14(2):101–114.
6. Lampe MA, Burlingame AL, Whitney J, Williams ML, Brown BE, Roitman E et al. Human stratum corneum lipids: Characterization and regional variations. *J Lipid Res*. 1983 Feb 1, 1983;24(2):120–130.

7. Prausnitz MR, Mitragotri S, Langer R. Current status and future potential of transdermal drug delivery. *Nat Rev Drug Discov.* 2004;3(2):115–124.
8. Illel B, Schaefer H, Wepierre J, Doucet O. Follicles play an important role in percutaneous absorption. *J Pharm Sci.* 1991 May;80(5):424–427.
9. Pirot F, Kalia YN, Stinchcomb AL, Keating G, Bunge A, Guy RH. Characterization of the permeability barrier of human skin in vivo. *Proc Natl Acad Sci USA.* 1997 Feb 18;94(4):1562–1567.
10. Masters BR, So PT, Gratton E. Multiphoton excitation fluorescence microscopy and spectroscopy of in vivo human skin. *Biophys J.* 1997;72(6):2405–2412.
11. Bernard E, Dubois J, Wepierre J. Importance of sebaceous glands in cutaneous penetration of an antiandrogen: Target effect of liposomes. *Am Chem Soc Am Pharmacol Assoc.* 1997;86(5):573–578.
12. Roberts MS, Cross SE. Dermal blood flow, lymphatics, and binding as determinants of topical absorption, clearance, and distribution. In: Riviere JE, editor. *Dermal Absorption Models in Toxicology and Pharmacology.* Boca Raton, FL: CRC/Taylor & Francis; 2006. pp. 251–281.
13. Scheuplein RJ, Blank IH. Permeability of the skin. *Physiol Rev.* 1971 Oct;51(4):702–747.
14. Roberts MS, Anissimov YG. Mathematical models in percutaneous absorption. In: Bronaugh RL, Maibach HI, editors. *Percutaneous Absorption Drugs—Cosmetics—Mechanisms—Methodology.* 4th edn. New York: Marcel Dekker; 2005. pp. 1–44.
15. Squier CA. The permeability of keratinised and nonkeratinised oral apithelium to horseradish peroxidase. *J Ultrastruct Res.* 1973;43:160–177.
16. Elias PM, Friend DS. The permeability barrier in mammalian epidermis. *J Cell Biol.* 1975 Apr;65(1):180–191.
17. Chandrasekaran SK, Shaw JE. Factors affecting the percutaneous absorption of drugs. *Curr Probl Dermatol.* 1975;7:142–155.
18. Elias PM, Goerke J, Friend DS. Mammalian epidermal barrier layer lipids: Composition and influence on structure. *J Invest Dermatol.* 1977 Dec;69(6):535–546.
19. Elias PM, Brown BE, Fritsch P, Goerke J, Gray GM, White RJ. Localization and composition of lipids in neonatal mouse stratum granulosum and stratum corneum. *J Invest Dermatol.* 1979 Nov;73(5):339–348.
20. Berenson GS, Burch GE. Studies of diffusion of water through dead human skin; the effect of different environmental states and of chemical alterations of the epidermis. *Am J Trop Med Hyg.* 1951 Nov;31(6):842–853.
21. Niemanic MK, Elias PM. In situ precipitation in novel cytochemical technique for visualization of permeability pathways in mammalian stratum corneum. *J Histochem Cytochem.* 1980;28:573–578.
22. Scheuplein RJ. Mechanism of percutaneous adsorption. I. Routes of penetration and the influence of solubility. *J Invest Dermatol.* 1965 Nov;45(5):334–346.
23. Barry BW. Drug delivery routes in skin: A novel approach. *Adv Drug Deliver Rev.* 2002 Nov 1;54:S31–S40.
24. Meidan VM, Bonner MC, Michniak BB. Transfollicular drug delivery—Is it a reality? *Int J Pharm.* 2005;306(1–2):1–14.
25. Scheuplein RJ. Mechanism of percutaneous absorption. II. Transient diffusion and the relative importance of various routes of skin penetration. *J Invest Dermatol.* 1967;48(1):79–88.
26. Scheuplein RJ, Blank IH. Permeability of the skin. *Physiol Rev.* 1971 Oct 1;51(4):702–747.
27. Grams YY, Whitehead L, Cornwell P, Bouwstra JA. Time and depth resolved visualization of the diffusion of a lipophilic dye into the hair follicle of fresh unfixed human scalp skin. *J Control Release.* 2004 Aug 27;98(3):367–378.
28. Grams YY, Alaruikka S, Lashley L, Caussin J, Whitehead L, Bouwstra JA. Permeant lipophilicity and vehicle composition influence accumulation of dyes in hair follicles of human skin. *Eur J Phar Sci.* 2003;18(5):329–336.

29. Vogt A, Combadiere B, Hadam S, Stieler KM, Lademann J, Schaefer H et al. 40 nm, but not 750 or 1,500 nm, nanoparticles enter epidermal CD1a+ cells after transcutaneous application on human skin. *J Invest Dermatol.* 2006 Jun;126(6):1316–1322.

30. Lademann J, Knorr F, Richter H, Blume-Peytavi U, Vogt A, Antoniou C et al. Hair follicles—An efficient storage and penetration pathway for topically applied substances. *Skin Pharm Physiol.* 2008;21:150–155.

31. Toll R, Jacobi U, Richter H, Lademann J, Schaefer H, Blume-Peytavi U. Penetration profile of microspheres in follicular targeting of terminal hair follicles. *J Invest Dermatol.* 2004 Jul;123(1):168–176.

32. Lademann J, Richter H, Jacobi U, Patzelt A, Hueber-Becker F, Ribaud C et al. Human percutaneous absorption of a direct hair dye comparing in vitro and in vivo results: Implications for safety assessment and animal testing. *Food Chem Toxicol.* 2008 Jun;46(6):2214–2223.

33. Ananthapadmanabhan KP, Yu KK, Meyers CL, Aronson MP. Binding of surfactants to stratum corneum. *J Soc Cosmet Chem.* 1996;47:185–200.

34. Aki H, Kawasaki Y. Thermodynamic clarification of interaction between antiseptic compounds and lipids consisting of stratum corneum. *Thermochim Acta.* 2004;416(1–2): 113–119.

35. Dancik Y, Thörling C, Krishnan G, Roberts MS. Cutaneous metabolism and active transport in transdermal drug delivery. In: Monteiro-Riviere N, editor. *Toxicology of the Skin.* Boca Raton, FL: Taylor & Francis Group LLC; 2010.

36. Hotchkiss SAM. Dermal metabolism. In: Roberts MS, Walters KA, editors. *Dermal Absorption and Toxicity Assessment.* Marcel Dekker, New York. 1998. pp. 43–101.

37. Wester RC, Christoffel J, Hartway T, Poblete N, Maibach HI, Forsell J. Human cadaver skin viability for in vitro percutaneous absorption: Storage and detrimental effects of heat-separation and freezing. *Pharm Res.* 1998 Jan;15(1):82–84.

38. Nakashima E, Noonan PK, Benet LZ. Transdermal bioavailability and first-pass skin metabolism: A preliminary evaluation with nitroglycerin. *J Pharmacokinet Biopharm.* 1987 Aug;15(4):423–437.

39. Cross SE, Roberts MS. Subcutaneous absorption kinetics and local tissue distribution of interferon and other solutes. *J Pharm Pharmacol.* 1993 Jul;45(7):606–609.

40. Megwa SA, Benson HA, Roberts MS. Percutaneous absorption of salicylates from some commercially available topical products containing methyl salicylate or salicylate salts in rats. *J Pharm Pharmacol.* 1995 Nov;47(11):891–896.

41. Marks R. Pharmacokinetics and safety review of tazarotene. *J Am Acad Dermatol.* 1998 Oct;39(4 Pt 2):S134–S138.

42. Tang-Liu DD, Matsumoto RM, Usansky JI. Clinical pharmacokinetics and drug metabolism of tazarotene: A novel topical treatment for acne and psoriasis. *Clin Pharmacokinet.* 1999 Oct;37(4):273–287.

43. Bronaugh RL, Kraeling EK, Yourick JJ, Hood HL. Cutaneous metabolism during in vitro percutaneous absorption. In: Bronaugh RL, Maibach HI, editors. *Percutaneous Absorption: Drugs Cosmetics Mechanisms Methodology,* 3rd edn. New York: Marcel Dekker; 1999. pp. 57–64.

44. Ando HY, Ho NF, Higuchi WI. Skin as an active metabolizing barrier I: Theoretical analysis of topical bioavailability. *J Pharm Sci.* 1977 Nov;66(11):1525–1528.

45. Yu CD, Fox JL, Ho NF, Higuchi WI. Physical model evaluation of topical prodrug delivery-simultaneous transport and bioconversion of vidarabine-5'-valerate II: Parameter determinations. *J Pharm Sci.* 1979 Nov;68(11):1347–1357.

46. Yu CD, Fox JL, Ho NF, Higuchi WI. Physical model evaluation of topical prodrug delivery—Simultaneous transport and bioconversion of vidarabine-5'-valerate I: Physical model development. *J Pharm Sci.* 1979 Nov;68(11):1341–1346.

47. Yu CD, Gordon NA, Fox JL, Higuchi WI, Ho NF. Physical model evaluation of topical prodrug delivery—Simultaneous transport and bioconversion of vidarabine-5'-valerate V: Mechanistic analysis of influence of nonhomogeneous enzyme distributions in hairless mouse skin. *J Pharm Sci*. 1980 Jul;69(7):775–780.

48. Yu CD, Fox JL, Higuchi WI, Ho NF. Physical model evaluation of topical prodrug delivery—Simultaneous transport and bioconversion of vidarabine-5'-valerate IV: Distribution of esterase and deaminase enzymes in hairless mouse skin. *J Pharm Sci*. 1980 Jul;69(7):772–775.

49. Yu CD, Higuchi WI, Ho NF, Fox JL, Flynn GL. Physical model evaluation of topical prodrug delivery—Simultaneous transport and bioconversion of vidarabine-5'-valerate III: Permeability differences of vidarabine and n-pentanol in components of hairless mouse skin. *J Pharm Sci*. 1980 Jul;69(7):770–772.

50. Liu P, Higuchi WI, Song WQ, Kurihara-Bergstrom T, Good WR. Quantitative evaluation of ethanol effects on diffusion and metabolism of beta-estradiol in hairless mouse skin. *Pharm Res*. 1991 Jul;8(7):865–872.

51. Liu P, Higuchi WI, Ghanem AH, Good WR. Transport of beta-estradiol in freshly excised human skin in vitro: Diffusion and metabolism in each skin layer. *Pharm Res*. 1994 Dec;11(12):1777–1784.

52. Sugibayashi K, Hayashi T, Morimoto Y. Simultaneous transport and metabolism of ethyl nicotinate in hairless rat skin after its topical application: The effect of enzyme distribution in skin. *J Control Release*. 1999 Nov 1;62(1–2):201–208.

53. Gysler A, Kleuser B, Sippl W, Lange K, Korting HC, Holtje HD. Skin penetration and metabolism of topical glucocorticoids in reconstructed epidermis and in excised human skin. *Pharm Res*. 1999 Sep;16(9):1386–1391.

54. Hatanaka T, Rittirod T, Katayama K, Koizumi T. Influence of enzyme distribution and diffusion on permeation profile of prodrug through viable skin: Theoretical aspects for several steady-state fluxes in two transport directions. *Biol Pharm Bull*. 1999 Jun;22(6):623–626.

55. Rittirod T, Hatanaka T, Uraki A, Hino K, Katayama K, Koizumi T. Species difference in simultaneous transport and metabolism of ethyl nicotinate in skin. *Int J Pharm*. 1999 Feb 15;178(2):161–169.

56. Seko N, Bando H, Lim CW, Yamashita F, Hashida M. Theoretical analysis of the effect of cutaneous metabolism on skin permeation of parabens based on a two-layer skin diffusion/metabolism model. *Biol Pharm Bull*. 1999 Mar;22(3):281–287.

57. Cross SE, Anderson C, Roberts MS. Topical penetration of commercial salicylate esters and salts using human isolated skin and clinical microdialysis studies. *Br J Clin Pharmacol*. 1998 Jul;46(1):29–35.

58. Roberts MS, Cross SE, Anissimov YG. Factors affecting the formation of a skin reservoir for topically applied solutes. *Skin Pharmacol Appl Skin Physiol*. 2004;17:3–16.

59. Magnusson BM, Cross SE, Winckle G, Roberts MS. Percutaneous absorption of steroids: Determination of in vitro permeability and tissue reservoir characteristics in human skin layers. *Skin Pharmacol Physiol*. 2006;19(6):336–342.

60. Potts RO, Guy RH. Predicting skin permeability. *Pharm Res*. 1992;9(5):663–669.

61. Magnusson BM, Anissimov YG, Cross SE, Roberts MS. Molecular size as the main determinant of solute maximum flux across the skin. *J Invest Dermatol*. 2004;122(4):993–999.

62. Zhang Q, Grice JE, Li P, Jepps OG, Wang GJ, Roberts MS. Skin solubility determines maximum transepidermal flux for similar size molecules. *Pharm Res*. 2009 Aug;26(8):1974–1985.

63. Riegelman S. Pharmacokinetic factors affecting epidermal penetration and percutaneous adsorption. *Clin Pharmacol Ther*. 1974 Nov;16(5 Pt 2): 873–883.

64. Roberts MS, Walters KA. Human skin morphology and dermal absorption. In: Roberts MS, Walters KA, editors. *Dermal Absorption and Toxicity Assessment*. New York: Informa Healthcare; 2009. pp. 1–15.

2 Theoretical Models for Dermatokinetics of Therapeutic Agents

Rong Shi and Hartmut Derendorf

CONTENTS

2.1 INTRODUCTION TO FUNDAMENTALS OF PHARMACOKINETICS

Pharmacokinetics is defined as the discipline to study the time course of drug and metabolite concentrations in the blood or tissue after drug administration. Pharmacokinetics can be used in drug development to optimize therapy; it is used to determine the dose, dosage regimen (the frequency of dosing), and dosage formulations. Choosing the wrong dose, dosage regimen, or dosage form will render any drug ineffective. It is important to not only pick the right compound but also the right dose, dosage regimen, or dosage form.

After the drug is administered, the "fate of the drug" will undergo liberation, absorption, distribution, metabolism, and, finally, excretion (LADME). Liberation is the release of the drug from the dosage form and can be important, for example, sustained release dosage forms. Absorption is the process of the drug entering the body. Compounds that do not have good absorption are not good candidates for oral dosage forms. The goal of pharmacokinetic optimization is to produce a desired drug concentration in the body using optimum dosage regimen and dosage form. Distribution is the process of the drug reaching the site of interest in the body where it can be active. Metabolism is the chemical breakdown of the drug molecules; the activity, toxicity, and half-life of the metabolites can be critical for the pharmacokinetics of the drug. Excretion is about the route of elimination of the drug and rate of elimination.

The LADME information can be used to find the current target concentration, but, for many drugs, the optimum target concentration may still need to be determined. The field that compares and correlates the concentration and the effect is called pharmacokinetics and pharmacodynamics (PKPD). The clinical pharmacokinetics concept is to study the pharmacokinetics in "normal," healthy subjects, to examine the biological variability between the subjects, and to monitor the therapeutic drug level in individual patients. Determination of pharmacokinetic parameters in "normal" subjects include measuring the drug and metabolite concentrations in the blood, urine, and other body fluids as a function of time; setting up models to describe the time course and predicting the drug levels as times not studied; and determining parameters from models related to the physiological, clinical, physical, or chemical properties of the drug.

2.1.1 PHARMACOKINETIC PARAMETERS

Pharmacokinetics uses mathematical modeling to quantify the relationship between dose and drug disposition. Pharmacokinetics and pharmacodynamics are closely related. Pharmacokinetics describes what the body does to the drug, while pharmacodynamics describes what the drug does to the body. Only when the relationship between pharmacokinetics and pharmacodynamics is studied will the pharmacokinetic work have values in clinicals. PKPD assumes that there is a correlation between the drug concentration in the systemic circulation and the pharmacological or toxic response, and that the plasma drug concentration is related to the drug concentration at the site of action. Various parameters are used to describe different pharmacokinetic processes. The key pharmacokinetic parameters are clearance, volume of distribution, half-life, bioavailability, and protein binding.

2.1.1.1 Clearance

Clearance quantifies the elimination. It is defined as the volume of body fluid cleared per time unit (L/h, mL/min), which is usually constant, and can be calculated as the rate of drug elimination to the plasma drug concentration or the rate of dose to the area under the concentration curve (AUC):

$$\text{Clearance} = \frac{\text{rate of elimination}}{\text{concentration or clearance}} = \frac{\text{dose}}{\text{area under the curve } (AUC)}$$

The concentration here is the drug concentration at steady state. When the elimination process is not saturated, the clearance is constant over a concentration range. When the fraction of the drug eliminated is constant per unit time, the elimination follows first-order kinetics. When the amount of drug eliminated is constant per unit time, the elimination follows zero-order kinetics. Ethanol, for example, follows zero-order kinetics. A lot of drugs follow first-order kinetics, unless the elimination or biotransformation is saturated, in which case the kinetics become zero order and clearance is not a constant any more.

The models of clearance are described as follows:

The well-stirred model for clearance is shown in Figure 2.1:

$$E = \frac{C_i - C_o}{C_i}$$

$$CL = QE$$

$$CL = \frac{Q f_u CL_{int}}{Q + f_u CL_{int}}$$

where

Q is the blood flow
CL_{int} is the intrinsic clearance
E is the extraction ratio
f_u is the fraction unbound to protein
C_i is the concentration of drug in the blood entering the organ
C_o is the concentration of drug in the blood leaving the organ

The extraction ratio is derived from $(C_i - C_o)/C_i$, and the ratio always falls between 0 and 1. When the extraction ratio is close to 0, the drug is referred to as a low-extraction drug, and when it is close to 1, it is a high-extraction drug.

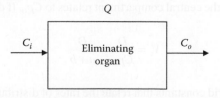

FIGURE 2.1 Well-stirred model for clearance.

Blood flow, intrinsic clearance, and protein binding provide information for the prediction of changes in clearance during steady state. For high-extraction drugs, $Q \ll f_u CL_{int}$, so $CL = Q$, and for low-extraction drugs, $Q \gg f_u CL_{int}$, so $CL = f_u CL_{int}$. The total body clearance is the sum of the individual organ clearances, for example, $CL = CL_{ren} + CL_{hep} + CL_{other}$.

2.1.1.2   Volume of Distribution

The volume of distribution, often called "apparent volume of distribution (Vd)", does not necessarily indicate a defined physiological space. It is simply the volume of compartment fluid required to account for the total amount of drug in the body at the same concentration as found in the blood or plasma. The volume of distribution quantifies distribution; it relates drug concentration (Cp) to the amount of drug in the body (X) and gives information on the amount of drug distributed into the tissues after the distribution equilibrium has been reached between the drug in the plasma and the tissue. In a one-compartment model, where the body is considered as one homogeneous unit, the volume of distribution can be expressed as

$$Vd = \frac{\text{amount of drug in the body}}{\text{plasma drug concentration}}$$

The volume of distribution can be seen as a proportionality constant between the amount of drug and plasma concentration. In this case, the drug is distributed instantaneously through the whole body right after the drug administration into the central compartment, oftentimes, through intravenous (IV) route. A linear pharmacokinetics is assumed in this case, where the elimination follows first-order kinetics, so that a change in dose will be reflected by a proportional change in the plasma concentration. When the apparent volume of distribution is larger than the plasma volume ($3 L$), the drug is present in the tissue or fluids beside the plasma compartment. A large volume of distribution indicates that the drug is widely distributed to the tissues. The volume of distribution depends on the plasma and tissue protein binding, lipophilicity, and pK_a of the drug.

However, very few drugs show immediate distribution and equilibrium through the body. For a lot of drugs, a one-compartment model does not explain the time course of distribution and elimination of the drug in the body. Therefore, a two-compartment model is a better choice for these drugs. In a two-compartment model with both central and peripheral compartments, there are three volumes of distribution terms: V_c (volume of distribution of central compartment), Vd_{ss} (volume of distribution at steady state), and Vd_{area} (volume of distribution at the elimination phase). Since the drug is present only in the central compartment at time $t = 0$, no distribution has occurred at this stage, so the volume distribution in the central compartment relates to Cp_0. If dose (D) is given, then

$$V_c = \frac{D}{Cp_0} = \frac{D}{a+b}$$

where a and b are hybrid constants that relate the rates of distribution and elimination, respectively.

At steady state, equilibrium is reached between the free plasma concentration and free tissue concentration. The volume of distribution at this point is called the volume of distribution at steady-state Vd_{ss}, which stands for the sum of volume terms of peripheral and central compartments at steady state.

$$Vd_{ss} = \frac{X_{ss}}{Cp_{ss}} = V_c + V_p = \left(1 + \frac{k_{12}}{k_{21}}\right)V_c$$

where k_{12} and k_{21} are the microconstants used in two-compartment models.

At the terminal elimination phase, the drug concentration in the peripheral compartment is in a dynamic equilibrium with the central compartment. Here, Vd_{area} relates to the amount of drug in the body and to the plasma concentration:

$$Vd_{area} = \frac{D}{AUC_\infty \beta}$$

where
AUC_∞ is the total area under the curve
β is the terminal elimination rate constant

Figure 2.2 shows the volume terms for these three phases, respectively; X_c indicates the central compartment and X_p indicates the peripheral compartment. The darker the color of the compartment, the greater is the amount of drug present in that compartment. To summarize, when comparing the three volume terms, $Vd_{area} > Vd_{ss} > V_c$.

2.1.1.3 Half-Life

Half-life is the time it takes for the concentration to fall to half of its previous value. It is important to remember that half-life is a secondary pharmacokinetics parameter and depends on the clearance and volume of distribution. As a result, changes to the clearance and volume of distribution, aging, drug interactions, etc. would lead to a change in half-life. For first-order linear pharmacokinetics, half-life is a more

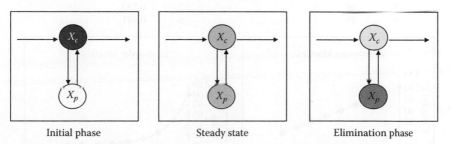

| Initial phase | Steady state | Elimination phase |

FIGURE 2.2 Distribution at three phases for a two-compartment model corresponding to three volume terms. (From Derendorf, H. et al, *Pharmakokinetik*. eds: Derendorf, H., Gramatté, T., Schäfer, G. Stuttgart, Germany: Wissenschaftliche Verlagsgesellschaft mbH, 2002.)

meaningful parameter than the elimination rate constant k_e. Half-life is typically determined from the log-terminal phase of the elimination curve:

$$t_{1/2} = \frac{\ln 2}{k_e} = \frac{0.693}{k_e} = \frac{0.693\,Vd}{CL}$$

where k_e is the elimination rate constant in the terminal phase.

For zero-order kinetics, the amount of drug eliminated always stays the same and the time it takes for the concentration to fall to 50% of its previous value changes; hence, the concept of half-life is no longer meaningful. The difference between zero- and first-order kinetics is shown in Figure 2.3. For zero-order kinetics, the amount of drug eliminated per time unit is the same, and the rate of elimination is independent of the amount of drug in the body. For first-order kinetics, the actual amount of drug eliminated per time unit is changing; however, the fraction of drug eliminated per time period is constant, and this constant term is called the elimination rate constant k_e. In first-order kinetics, the rate of elimination is dependent on the amount of drug; the greater the amount of drug in the body, the more the drug is eliminated. The elimination rate constant k_e can be calculated given the concentrations and the time points those concentration samples are taken, by using the following equation:

$$k_e = \frac{CL}{Vd} = \frac{\ln\left(\dfrac{Cp_1}{Cp_2}\right)}{(t_2 - t_1)} = \frac{\ln Cp_1 - \ln Cp_2}{(t_2 - t_1)}$$

Zero-order kinetics		First-order kinetics	
t (h)	C (ng/mL)	t (h)	C (ng/mL)
0	100	0	100
1	90	1	50
2	80 $\quad C = C_0 - k_0{}^*t$	2	25 $\quad C = C_0{}^*\exp(-k_0{}^*t)$
3	70	3	12.5 $\quad k_e = 0.693\ 1/h$
4	60	4	6.25
5	50	5	3.125
6	40	6	1.563
7	30	7	0.782
8	20	8	0.391

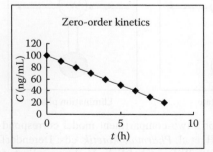

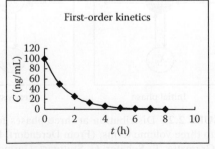

FIGURE 2.3 Comparison of zero- and first-order kinetics.

2.1.1.4 Bioavailability

After a certain dose of drug is given, how much of that dose will reach the systemic circulation is an important question in pharmacokinetics. Bioavailability (F) quantifies absorption; it is the fraction of dose that reaches the systemic circulation. Bioavailability discusses both the rate and the extent of absorption at the site of drug administration.

For drugs administered intravenously, the access to the systemic circulation is gained immediately after the administration; in this case, 100% bioavailability is reached. In the case of oral administration, bioavailability can be calculated by the following equation if the hepatic extraction ratio is known:

$$F = 1 - E = 1 - \left(\frac{CL_{hepatic}}{Q_{hepatic}} \right)$$

Oftentimes, the bioavailability of a drug via various administration routes (i.e., oral, transdermal, or by inhalation) can be determined by comparing the AUC of the plasma concentration after the administration and the AUC of plasma concentration after IV administration, which serves as a reference, for example, oral bioavailability:

$$F = \frac{AUC_{oral}}{AUC_{IV}}$$

Bioequivalence is determined by showing that no statistical difference exists among C_{max}, t_{max}, and AUC for the generic and reference drugs.

The reasons for poor oral bioavailability can be (1) poor solubility: drug molecules have to be in the solution before crossing the membrane; a low solubility leads to poor bioavailability; (2) first-pass effect in the intestinal wall: during the absorption of the drug molecules from the intestine to the blood, the enzymes in the intestinal wall can lead to degradation of the molecules; (3) first-pass effect in the liver: before the drug can reach the systemic circulation the first time, it has to go through the liver; metabolism enzymes can metabolize the drug, and transporter proteins, the so-called efflux transporters (e.g., the most important and well-studied P-glycoprotein), can remove the drug back into the intestinal lube; (4) small absorption windows: uptake of the drug into the blood has to be between the absorption window: falling outside a small absorption window will also result in a low bioavailability; (5) large molecular size: large molecules, such as large proteins, which have difficulty in passing through the membrane, will lead to a low absorption; and (6) instability in the gastrointestinal (GI) tract: oftentimes, proteins are not stable in the GI tract, for example, insulin cannot be administered orally, because the proteases in the GI tract will break down the molecule before the drug is absorbed. Other factors can also have influences on bioavailability such as the salt factor and the dosing rate. For two drug absorption profiles that have the same extent but different rates, the faster the absorption, the higher the maximum concentration, and the earlier the t_{max}.

2.1.1.5 Protein Binding

Protein binding is an important factor in pharmacokinetics because it is the free drug in the plasma and tissue that contribute to the therapeutic effect. The free (unbound) concentration of the drug at the receptor site should be used in PK/PD correlations to make predictions for pharmacological activity. Drug binds to protein in a reversible fashion; the binding is not a covalent bond but a charge interaction between the protein and the drug molecule. After the free drug is consumed, previously bound drug molecules are released. This quick equilibrium of binding occurs in milliseconds. The protein binding normally follows linear kinetics because, oftentimes, we have an unlimited number of proteins that are independent of the concentration of the drug. If the proteins are saturated, which is only for a few drugs, the binding then no longer follows first-order kinetics. There are a few methods to determine the protein binding: ultrafiltration, ultracentrifugation, equilibrium dialysis, microdialysis (MD), erythrocyte-binding method, etc. For high-extraction drugs, the plasma concentration will not be affected by protein binding, but the free plasma concentration will be; for low-extraction drugs, protein binding will have an impact on the free plasma concentration but not on the whole plasma concentration.

The protein binding (fraction unbound f_u) can have an impact on parameters such as Vd, CL, and F.

The relation between protein binding and Vd can be expressed as

$$Vd = V_p + \frac{f_u}{f_{uT}} \times V_T$$

Clearance can be calculated as

$$CL_{hepatic} = \frac{Q_{hepatic} \times f_u \times CL_{int}}{Q_{hepatic} + f_u \times CL_{int}} = Q_{hepatic} \times E$$

For low-extraction drugs, protein binding changes lead to clearance changes. The above equation can be simplified as

$$CL_{hepatic} = f_u \times CL_{int}$$

For high-extraction drugs, protein binding changes do not lead to changes in clearance:

$$CL_{hepatic} = Q_{hepatic}$$

Bioavailability with protein binding:

$$F_H = 1 - E = \frac{Q_{hepatic}}{Q_{hepatic} + f_u \times CL_{int}}$$

For high-extraction drugs, changes and protein binding will lead to changes in bioavailability:

$$F_{H,high} = 1 - E = \frac{Q_{hepatic}}{f_u \times CL_{int}}$$

For low-extraction drugs, $F \approx 1$, protein binding changes do not lead to changes in bioavailability.

2.1.2 COMPARTMENT MODELS IN PHARMACOKINETICS

2.1.2.1 One-Compartment Model

2.1.2.1.1 Intravenous Bolus One-Compartment Model

The simplest one-compartment body model is the model with IV bolus administration (Figure 2.4), where X is the amount of drug in the body, E is the amount of drug eliminated from the body, the input is IV bolus into the body, B is the amount of drug eliminated via bile excretion, U is the amount of drug eliminated by urinary excretion, M is the amount of drug metabolized, and U^M is the amount of metabolites eliminated. It has a first-order elimination rate constant k_e, where k_e is the secondary parameter related to the clearance and volume of distribution. Rate constants k_B, k_R, k_M, and k_e^M are the elimination rate constants for the parallel elimination routes, correspondingly. The plasma concentration in the model can be described by the equation with Cp_0 as the initial concentration after IV administration:

$$Cp = Cp_0 e^{-k_e t} \quad Cp_0 = \frac{D}{Vd}$$

The plasma concentration and time profile is shown in Figure 2.5 with both a normal scale plot and a semilogarithmic plot.

It is also important to study the pharmacokinetics of the metabolites of the compounds since metabolites often have efficacy or toxicity properties. There are often

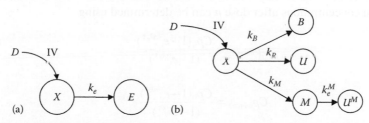

FIGURE 2.4 (a) One-compartment model with IV bolus administration; (b) one-compartment model with IV bolus administration and parallel elimination. (From Derendorf, H. et al, *Pharmakokinetik*. eds: Derendorf, H., Gramatté, T., Schäfer, G. Stuttgart, Germany: Wissenschaftliche Verlagsgesellschaft mbH, 2002.)

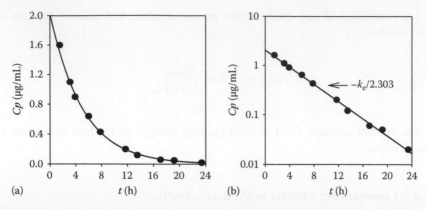

FIGURE 2.5 Plasma concentration and time profile for one-compartment model with single-dose IV bolus administration. (a) Normal plot. (b) Semilogarithmic plot. (From Derendorf, H. et al, *Pharmakokinetik*. eds: Derendorf, H., Gramatté, T., Schäfer, G. Stuttgart, Germany: Wissenschaftliche Verlagsgesellschaft mbH, 2002.)

two types of situation for metabolites (Figure 2.6): elimination-limited metabolite pharmacokinetics and formation-limited metabolite pharmacokinetics.

For the elimination-limited metabolite pharmacokinetics, the metabolite has a longer half-life than the parent drug; the actual half-life of the parent drug profile will be the same as if there is only the metabolite itself administered. For the formation-limited metabolite pharmacokinetics, the plasma concentration and time profile of the parent drug (alprenolol in Figure 2.6b) and the metabolite (4-hydroxyalprenolol) deplete in parallel. However, by itself, the metabolite has a much shorter half-life, for example, 4-hydroxyalprenolol (IV), shown in Figure 2.6b. The metabolite and the parent drug have the same half-life because the elimination of the metabolite is limited by the formation of the metabolite from the parent drug.

By adding multiple new doses to the existing single dose at certain dosing intervals, it gives a profile of multiple-dose administration. Usually, for the first four or five dosing intervals (τ), the peak concentration (C_{max}) and the lowest concentration (C_{min}, also called trough concentration) increase until a steady state is reached (Figure 2.7). The distance from peak and trough concentrations remains constant, which is called steady-state fluctuation. In theory, we would like both peak and trough concentrations to stay within the therapeutic window. The maximum and minimum concentrations after dose n can be determined using

$$Cp_{max(n)} = \frac{Cp_0(1 - e^{-nk_e\tau})}{(1 - e^{-k_e\tau})}$$

$$Cp_{min(n)} = \frac{Cp_0(1 - e^{-nk\tau})}{(1 - e^{-k\tau})} e^{-k\tau}$$

where

n is the nth dose

τ is the dosing interval

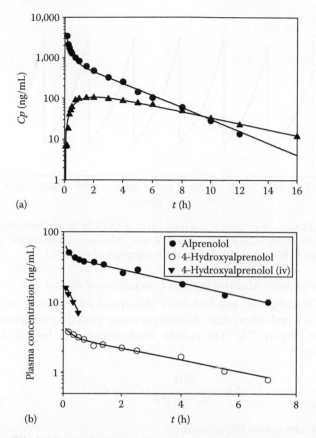

(a)

(b)

FIGURE 2.6 Elimination-limited and formation-limited metabolite pharmacokinetics. (a) Elimination-limited metabolite PK. (b) Formation-limited metabolite PK. (From Derendorf, H. et al, *Pharmakokinetik*. eds: Derendorf, H., Gramatté, T., Schäfer, G. Stuttgart, Germany: Wissenschaftliche Verlagsgesellschaft mbH, 2002.)

The maximum, minimum, and average concentrations of steady state for multiple IV doses are shown in the following equations:

$$Cp_{max(ss)} = \frac{Cp_0}{(1 - e^{-k_e \tau})}$$

$$Cp_{min(ss)} = \frac{Cp_0}{(1 - e^{-k_e \tau})} e^{-k_e \tau}$$

$$Cp_{ave(ss)} = \overline{C}p_{ss} = \frac{D}{CL\tau}$$

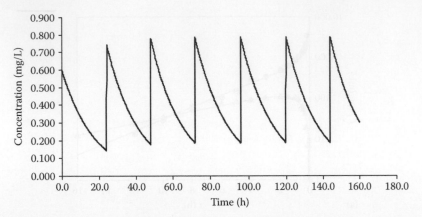

FIGURE 2.7 Plasma concentration and time curve with IV bolus multiple-dose administration. (From Derendorf, H. et al, *Pharmakokinetik*. eds: Derendorf, H., Gramatté, T., Schäfer, G. Stuttgart, Germany: Wissenschaftliche Verlagsgesellschaft mbH, 2002.)

2.1.2.1.2   First-Order Absorption One-Compartment Model

The fraction absorbed (*A*) and first-order absorption rate constant (k_a) are added in the model for oral absorption one-compartment model, as compared to the IV administration (Figure 2.8). The plasma concentration can be calculated by the Bateman equation

$$Cp = \frac{FDk_a}{(k_a - k_e)Vd}(e^{-k_e t} - e^{-k_a t})$$

where k_a is the absorption rate constant.

The plasma concentration and time profile is shown in Figure 2.9 with both a normal scale plot and a semilogarithmic plot.

The constants k_a and k_e can be determined by the method of residuals if $k_a \gg k_e$, as shown in Figure 2.10. Let $Cp' = (k_a FD/Vd(k_a - k_e))e^{-k_e t}$

The terminal slope using the above equation represents k_e if plotted in a semilogarithmic scale.

If we calculate $Cp' - Cp$, then

$$Cp' - Cp = \left(\frac{k_a FD}{Vd(k_a - k_e)}\right)e^{-k_a t}$$

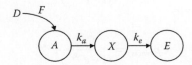

FIGURE 2.8 One-compartment model with oral administration. (From Derendorf, H. et al, *Pharmakokinetik*. eds: Derendorf, H., Gramatté, T., Schäfer, G. Stuttgart, Germany: Wissenschaftliche Verlagsgesellschaft mbH, 2002.)

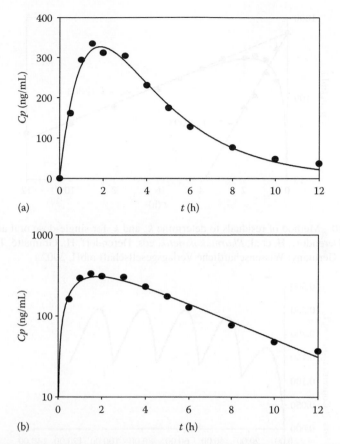

(a)

(b)

FIGURE 2.9 Plasma concentration and time profile for one-compartment model with single-dose oral administration. (a) Normal plot. (b) Semilogarithmic plot. (From Derendorf, H. et al, *Pharmakokinetik*. eds: Derendorf, H., Gramatté, T., Schäfer, G. Stuttgart, Germany: Wissenschaftliche Verlagsgesellschaft mbH, 2002.)

The terminal slope using the above equation represents k_a if plotted in a semilogarithmic scale.

The method of residuals can be used only when $k_a \gg k_e$. It cannot be applied when the rate of absorption is lower than the rate of elimination ($k_a < k_e$), that is, in a "flip-flop" system with absorption rate-limited elimination. In a "flip-flop" scenario, k_e depends on k_a.

For multiple-dose oral administration (Figure 2.11), the concentration after dose n can be calculated as

$$Cp = \frac{FDk_a}{(k_a - k_e)Vd}\left(\frac{(1 - e^{-nk_e\tau})e^{-k_e t}}{(1 - e^{-k_e\tau})} - \frac{(1 - e^{-nk_a\tau})e^{-k_a t}}{(1 - e^{-k_a\tau})}\right)$$

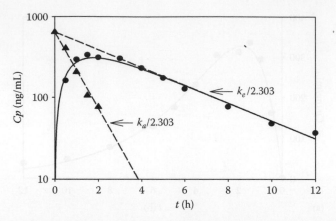

FIGURE 2.10 Method of residuals to determine k_a and k_e for single-dose oral administration. (From Derendorf, H. et al, *Pharmakokinetik*. eds: Derendorf, H., Gramatté, T., Schäfer, G. Stuttgart, Germany: Wissenschaftliche Verlagsgesellschaft mbH, 2002.)

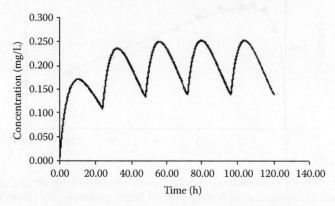

FIGURE 2.11 Plasma concentration and time profile with oral multiple-dose administration. (From Derendorf, H. et al, *Pharmakokinetik*. eds: Derendorf, H., Gramatté, T., Schäfer, G. Stuttgart, Germany: Wissenschaftliche Verlagsgesellschaft mbH, 2002.)

The plasma concentration at steady state can be calculated using the following equation:

$$Cp_{ss} = \frac{k_a FD}{Vd(k_a - k_e)} \left(\frac{e^{-k_e t}}{(1 - e^{-k_e \tau})} - \frac{e^{-k_a t}}{(1 - e^{-k_a \tau})} \right)$$

The average concentration at steady state can be calculated with the same equation for multiple IV doses.

2.1.2.1.3 One-Compartment Model for Infusion

For constant rate infusion, the plasma concentration increases gradually until it reaches the steady-state plateau (Figure 2.12). During steady state, the amount delivered is equal to the amount being eliminated from the body. After the infusion is stopped, plasma concentration decreases and this can be described by $e^{-k_e t}$ (Figure 2.13).

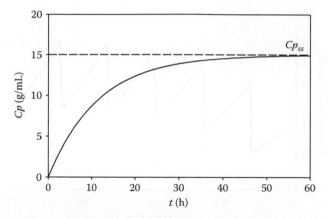

FIGURE 2.12 Plasma concentration and time profile with constant rate infusion. (From Derendorf, H. et al, *Pharmakokinetik*. eds: Derendorf, H., Gramatté, T., Schäfer, G. Stuttgart, Germany: Wissenschaftliche Verlagsgesellschaft mbH, 2002.)

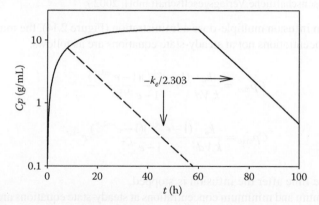

FIGURE 2.13 Plasma concentration and time profile after stop of constant rate infusion. (From Derendorf, H. et al, *Pharmakokinetik*. eds: Derendorf, H., Gramatté, T., Schäfer, G. Stuttgart, Germany: Wissenschaftliche Verlagsgesellschaft mbH, 2002.)

The plasma concentration for constant rate infusion can be calculated by the following equation:

$$Cp = \frac{R_0}{k_e Vd}(e^{k_e T} - 1)e^{-k_e t}$$

Plasma concentration during infusion and at steady state (Cp_{ss}) can be obtained by the following equations:

$$Cp = \frac{R_0}{k_e Vd}(1 - e^{-k_e t})$$

$$Cp_{ss} = \frac{R_0}{CL}$$

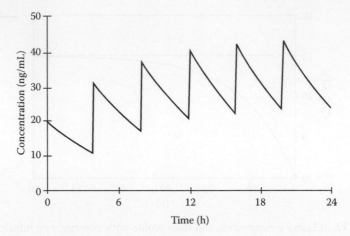

FIGURE 2.14 Plasma concentration and time profile with short-term infusion. (From Derendorf, H. et al, *Pharmakokinetik*. eds: Derendorf, H., Gramatté, T., Schäfer, G. Stuttgart, Germany: Wissenschaftliche Verlagsgesellschaft mbH, 2002.)

For short-term infusion multiple-dose administration (Figure 2.14), the maximum and minimum concentrations not at steady-state equations are as follows:

$$Cp_{max} = \frac{k_o}{k_e Vd} \frac{(1 - e^{-k_e t})(1 - e^{-nk_e \tau})}{1 - e^{-k_e \tau}}$$

$$Cp_{max} = \frac{k_o}{k_e Vd} \frac{(1 - e^{-k_e t})(1 - e^{-nk_e \tau})}{1 - e^{-k_e \tau}} e^{-k_e t'}$$

where t' is the time after the infusion is stopped.

The maximum and minimum concentrations at steady-state equations are as follows:

$$Cp_{max} = \frac{k_o}{k_e V_d} \frac{(1 - e^{-k_e t})}{1 - e^{-k_e \tau}}$$

$$Cp_{min} = \frac{k_o}{k_e V_d} \frac{(1 - e^{-k_e t})}{1 - e^{-k_e \tau}} e^{-k_e t'}$$

2.1.2.2 Two-Compartmental Model

After drug administration, there is a time period for the drug to distribute through the systemic circulation and the organs, which is usually called the distribution phase. Drug plasma concentration during this distribution phase usually decreases more dramatically than that during the terminal elimination phase. Drug is usually distributed by blood flow and to highly perfused organs such as the liver, kidneys, and lungs, which have rapid distribution equilibrium with the blood. In a two-compartment model, blood and highly perfused organs can be "lumped" together as a homogenous well-stirred compartment that is called central compartment. This does not necessarily mean that the concentrations in all the tissues are the same; rather, it assumes that plasma concentration changes reflect all the tissue concentration changes in the central compartment. Organs that are poorly perfused (e.g., muscle, lean tissue, and fat) can be "lumped" together as the peripheral compartment. Followed by an IV bolus injection, the drug levels in the central compartment decline more rapidly during the distribution phase (α-phase) than during the elimination phase (β-phase). The volume terms V_c, Vd_{ss}, and Vd_{area} in a two-compartment model have been discussed earlier. After administration, the drug first distributes in the central compartment (V_c) and then moves into the peripheral compartment because of the existing concentration gradient. When free levels in the central and peripheral compartments are the same, the net transport of the drug stops and a steady state is reached (Vd_{ss}), but elimination continues. This is why the drug concentration in the central compartment becomes lower (Figure 2.16). Drug molecules move from the peripheral compartment into the central compartment while elimination continues, and the drug levels decline parallel to each other in the two compartments (Figure 2.15).

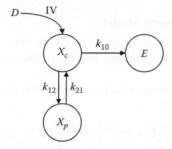

FIGURE 2.15 Two-compartment model with IV bolus administration. (From Derendorf, H. et al, *Pharmakokinetik*. eds: Derendorf, H., Gramatté, T., Schäfer, G. Stuttgart, Germany: Wissenschaftliche Verlagsgesellschaft mbH, 2002.)

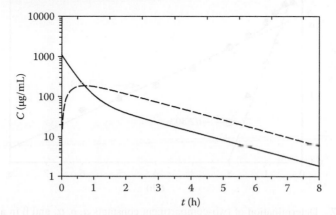

FIGURE 2.16 Two-compartment model plasma concentration and time profile in the central (solid line) and peripheral compartments (dashed line). (From Derendorf, H. et al, *Pharmakokinetik*. eds: Derendorf, H., Gramatté, T., Schäfer, G. Stuttgart, Germany: Wissenschaftliche Verlagsgesellschaft mbH, 2002.)

A two-compartment model has two exponential phrases in the equation, one for distribution (α-term) and the other for elimination (β-term):

$$Cp = ae^{-\alpha t} + be^{-\beta t}$$

where

a and b are hybrid constants that relate the rates of distribution and elimination, respectively

α and β are hybrid distribution rate constant and elimination rate constant corresponding to the α- and β-phases, respectively

The constants a, b, α, and β can be determined in a semilogarithmic plot (Figure 2.17).

The concentration–time profile for multiple-dose IV administration in a two-compartment model is shown in Figure 2.18, where $Cp_{ss(max)}$ and $Cp_{ss(min)}$ are the steady-state maximum and minimum concentrations, respectively.

2.1.2.3 Three-Compartment Model

A three-compartment model is shown in Figure 2.19, where X_p^s is the amount of drug in the shallow peripheral compartment and X_p^d is the amount of drug in the deep peripheral compartment.

The general concentration equation for a three-compartment model is

$$C = ae^{-\alpha t} + be^{-\beta t} + ce^{-\gamma t}$$

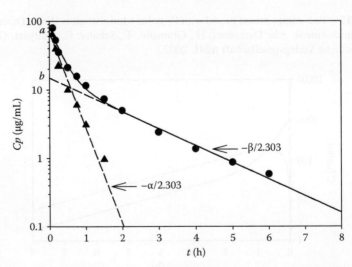

FIGURE 2.17 Determination of two-compartment constants a, b, α, and β in a semilogarithmic plot. (From Derendorf, H. et al, *Pharmakokinetik*. eds: Derendorf, H., Gramatté, T., Schäfer, G. Stuttgart, Germany: Wissenschaftliche Verlagsgesellschaft mbH, 2002.)

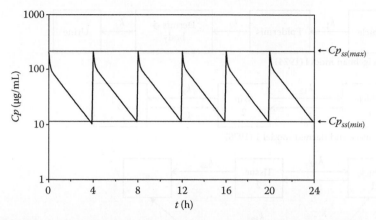

FIGURE 2.18   Concentration–time profile for multiple-dose IV administration in a two-compartment model. (From Derendorf, H. et al, *Pharmakokinetik*. eds: Derendorf, H., Gramatté, T., Schäfer, G. Stuttgart, Germany: Wissenschaftliche Verlagsgesellschaft mbH, 2002.)

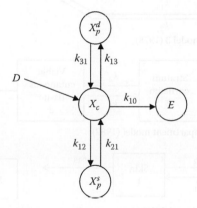

FIGURE 2.19   Three-compartment model. (From Derendorf, H. et al, *Pharmakokinetik*. eds: Derendorf, H., Gramatté, T., Schäfer, G. Stuttgart, Germany: Wissenschaftliche Verlagsgesellschaft mbH, 2002.)

where
  a, b, and c are hybrid constants that relate the rates of distribution and elimination, respectively
  α, β, and γ are the hybrid distribution rate, rapid, and slow elimination rate constants, respectively

The concentration and time profile after IV administration for a three-compartment model is shown in Figure 2.20.

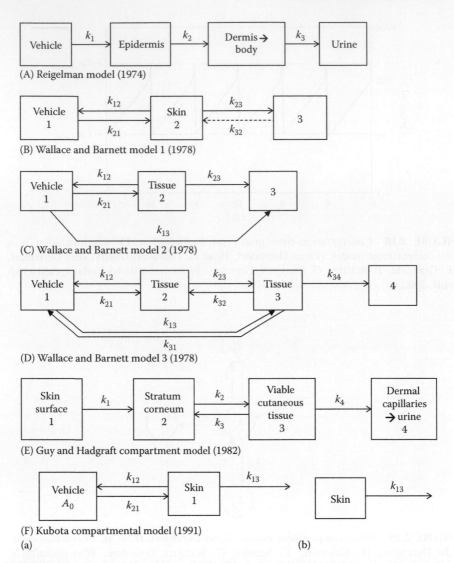

FIGURE 2.20   Concentration time profile for IV administration in a three-compartment model.

2.2   THEORETICAL MODELS OF DERMATOKINETICS OF DRUGS

Skin is the primary defense of our body from chemical invasion, and consists of three major layers: the stratum corneum (sc), the viable epidermis (ve), and the dermis (de); sc and ve together are usually called epidermis (epi). Chemicals absorbed first have to pass the outmost sc, which consists of dead, flattened, keratin-rich cells surrounded by an intracellular lipid-rich "mortar" (Michaels et al. 1975); as a result, it provides the main permeation resistance of most hydrophilic chemicals. The mass transfer of lipophilic chemicals through the sc is due to diffusion (Albery and Hadgraft 1979; el Tayar et al. 1991). Local equilibrium is often assumed between the outer layer of the sc and the vehicle that is often assumed to be well stirred. After penetration through the sc, chemicals enter the ve, which is composed of living cells, is relatively hydrophilic, and provides resistance mainly to lipophilic chemicals (Michaels et al. 1975). The de is highly vascularized and usually does not provide any significant resistance. The concentration gradient that leads to passive diffusion across both the sc and ve depends on the physicochemical and physical parameters of the chemicals and properties of the skin. These parameters are permeability coefficients P, partition coefficients K (e.g., sc–ve partition coefficients and skin–blood partition coefficients), cutaneous blood flow rate, diffusion coefficients D, and physical properties of the skin. The time required to diffuse across the skin layers, the lag time, also depends on these parameters.

Models in which skin is considered as a single homogeneous tank or diffusion membranes have been often used to analyze percutaneous absorption. The previously reported models are mostly diffusion membrane models and compartmental pharmacokinetic models. Although percutaneous absorption models of skin as one or two homogenous well-stirred tank reactors are generally used to describe a model, studies have shown that skin behaves like a membrane and membrane models of the skin appear to describe the skin nature more accurately (Scheuplein 1978). However, models treating the skin as a membrane also can be much more mathematically cumbersome. Meanwhile, compartmental models for percutaneous absorption have been developed and used as a simpler alternative. Compartment models have shown to provide easy incorporation into physiologically based pharmacokinetic models for chemicals (McCarley and Bunge 1998, 2000). This chapter mainly focuses on pharmacokinetic compartmental models developed for the percutaneous absorption of chemicals. Kalia and Guy (2001) described the theoretical principles and reviewed the mathematical models representing the drug release based on the formulation conditions. Roberts et al. (1999) summarized the mathematical models of in vitro percutaneous absorption for many boundary conditions related to the solutes crossing the membrane, as well as in vivo diffusion, compartmental, and physiologically based pharmacokinetics models. McCarley and Bunge reviewed pharmacokinetic compartment models for which rate constants were expressed in terms of the physicochemical and physical properties of the skin; 9 one-compartment and 2 two-compartment models were reviewed. The review (McCarley and Bunge 2001) summarized one-compartment models according to the properties of various skin layers (sc, ve, de), in terms of transport, resistance, and storage properties, by assuming whether particular layers are negligible or not. Two-compartment skin models are usually employed when neither the sc nor ve can be neglected because both the layers contribute to transport, resistance, or storage function of the skin. The contribution of this work has been a consistent nomenclature for the coefficients assigned to different compartment models, systematically derived equations for the coefficients of different compartment models, and all of which presented in terms of physiochemical and physical properties. After comparing 11 compartment models with the two-membrane model, the Kubota model, which includes both the sc and the ve, was considered by McCarley as the model that most closely estimates the two-membrane model for both hydrophilic and lipophilic chemicals, whereas the McCarley model, which includes just the sc, predicts only the hydrophilic chemicals well. Among the 11 one-compartment models reviewed in this chapter, a few common assumptions were used. In some models, the local equilibrium between the blood and skin or vehicle and skin were assumed; that is, the transfer rates across the skin are related to the skin/blood or vehicle/skin partition coefficient. It is commonly assumed that the rate of mass transfer into the skin from the vehicle is proportional to the steady-state permeability coefficient, and the clearance rate from the skin depends on the solubility of the compound and the cutaneous blood flow rate/area of the skin, which will lead to a skin concentration of zero if clearance from the skin by blood is faster than the transfer rate through the skin.

2.2.1 COMPARTMENT MODELS IN DERMAL ABSORPTION

Riegelman first proposed a concept of the rate process involved in the dermal permeation and percutaneous absorption (Riegelman 1974), a concept of unidirectional simple compartment-based pharmacokinetic model. Hostynek provided a correlation of in vitro and in vivo percutaneous absorption by a mathematic model (Hostynek et al. 1995).

Wallace and Barnett (1978) developed a series of pharmacokinetic compartment models to analyze the percutaneous absorption of methotrexate through full thickness hairless mouse skin, including models with mass transfer across the appendageal (i.e., shunt) pathway. A few assumptions were used in their models: first, the amount of drug in the vehicle is assumed constant for the suspension; second, skin was treated as a single homogeneous well-stirred barrier to penetration; and third, sink condition was assumed in the systematic circulation compartment (compartment 3) such that $k_{32} A_3$ is negligible. The sink condition is applied in the model because the concentration of the chemical in the circulation compartment is extraordinarily higher than the concentration of the chemical in the skin. The amount of chemical in the compartment 3 A_3, the diffusion rate or flux (J_s), and the lag time (τ) in the Wallace and Barnett model 1 are given by the following equations (Wallace and Barnett 1978):

$$A_3 = \frac{A_1 k_{12} k_{23}}{k_{12} + k_{23}} \left\{ t - \frac{1 - e^{[-t(k_{21} + k_{23})]}}{k_{21} + k_{23}} \right\} \tag{2.1}$$

The exponential term approaches zero as $t \rightarrow \infty$, and the above equation can be rearranged as

$$A_3 = \frac{A_1 k_{12} k_{23}}{k_{12} + k_{23}} \left(t - \frac{1}{k_{21} + k_{23}} \right) \tag{2.2}$$

$$J_s = \frac{A_1 k_{12} k_{23}}{E_2}$$

$$\tau = \frac{1}{E^2}$$

An alternative model (Wallace and Barnett model 2) consisting of a tissue compartment and a parallel shunt pathway was presented and this model showed an increased flux and a decreased lag time compared to Wallace and Barnett model 1. The shunt represents physiological, interappaendageal, and intercellular routes. Another model (Wallace and Barnett model 3) considering skin as a series of tissues, for example, sc, lower ep layer and de, was presented, along with the shunt pathway, but the model did not yield more information because of insufficient data to determine the exponential coefficients. A table of flux (J_s), lag time (τ), and exponential coefficient for Wallace and Barnett models 1 and 2 were summarized by the authors.

Guy and Hadgraft (1982) developed a model, and they first introduced k_{21} as a negligible parameter, such that a second sink condition was applied because the chemical concentration in the vehicle is extraordinarily larger than the chemical concentration in the skin. The solution for the fourth (urine) compartment is given by this model:

$$\phi_t = \frac{V_4 c_4}{V_1 c_0} = F k_1 k_2 k_4 \left\{ \frac{1}{k_1 \alpha \beta} - \frac{e^{-k_1 t}}{k_1 (k_1 - \alpha)(k_1 - \beta)} - \frac{e^{-\alpha t}}{\alpha(\alpha - \beta)(\alpha - k_1)} - \frac{e^{-\beta t}}{\beta(\beta - k_1)(\beta - \alpha)} \right\}$$

where

ϕ_t is the ratio of the amount of drug that has reached the urine at time t to the amount in compartment 1 at $t = 0$

F is the fraction of the applied topical dose plus metabolites recovered in the fourth (urine) compartment

V_i and c_i are, respectively, the volume of and the drug concentration in compartment i

c_0 is the concentration in compartment 1 at $t = 0$

Additionally, α and β are the roots of the following quadratic equation:

$$s^2 + (k_1 + k_2 + k_4)s + k_2 k_4 = 0$$

where k_1, k_2, k_3, and k_4 are the first-order rate constants shown in the Guy model.

The model by Guy et al. (1982) (Guy model) facilitated the in vivo modeling of urine concentration of three chemicals applied after topical application: testosterone, benzoic acid, and hydrocortisone. The four first-order kinetic processes were found to be related to the amount of drug excreted as a function of time, and the physical process of the rate constants was related to the rate constants in this model as well. Guy et al. (1983) used this model to describe the plasma concentrations after multiple topical doses of hydrocortisone after percutaneous absorption. In a later study, Guy et al. used the same model and applied it to 12 chemicals: aspirin, benzoic acid, benzyl nicotinate, caffeine, chloramphenicol, colchicine, dinitrochlorobenzene, diethyltoluamide, malathion, methyl nicotinate, nitrobenzene, and salicylic acid (Guy et al. 1985). In Guy models, the steady-state permeability coefficient is related to a first-order resistance in the vehicle and is described as the sc properties. The assumption in these models was that the chemical was considered as a solid on the surface of the skin, which led to the condition that cumulative amounts of chemical crossing vehicle and epidermis interface is assumed to be proportional to the mass of chemical remaining on the surface of the skin and the steady-state permeability coefficient. This model differs from most of the other models in that the steady-state permeability coefficient includes the surface area of the chemical instead of the volume of the vehicle. On comparing with membrane models, McCarley and Bunge (1998) concluded that the Guy model overestimates the mass transfer of chemicals across the epi and blood interface, which results in the chemical concentration in the vehicle dramatically decreasing when compared with two membrane models. Most of the compartment models assume that equilibrium between the skin and the vehicle are consistent at $t = 0$, but inconsistent over a long exposure period, and only the Kubota model predicts the same results as the two membrane models do over long exposure times.

Kubota and Maibach (1991) first developed rate constant expressions to make a one-compartment model match the properties of the membrane model. They proposed a compartment model for percutaneous chemical absorption where the skin–vehicle complex is represented by a two-compartment model with intercompartmental transfer rate constants (Kubota 1991):

$$T_{Lv} = \frac{1}{k_{12} + k_{13}} - \frac{1}{k_{13}}$$

$$T_{Ls} = \frac{1}{k_{12} + k_{13}}$$

$$J_{ss} = \left(\frac{k_{12}k_{13}}{k_{12} + k_{13}} \right) A_0$$

The above equations give the following equations when the three statistical moments are used as intercompartmental transfer rate constants, and which are defined as (1) the mean residence time (MRT) of a drug in the vehicle (MRT_v) and (2) in the skin (MRT_s), and (3) the variance of the residence time in the vehicle (VRT_v).

$$T_{Lv} = \frac{(MRT_v^2 - VRT_v)}{2MRT_v}$$

$$T_{Ls} = \frac{(MRT_v^2 - VRT_v)}{2MRT_v} + MRT_s$$

$$J_{ss} = \frac{A_0}{MRT_v}$$

The above compartment models in skin are summarized in Figure 2.21.

The percutaneous absorption model can be more complex when diffusion and metabolism both present at the same time. Guy and Hadgraft (1984) used their pharmacokinetic models (Figure 2.22), including a metabolic process k_5 and clearance of the metabolite k_6. Guy and Hadgraft (1982) and Kubota et al. (1995) used approximations to diffusion-based models to express the linear and saturable metabolism (Michaelis–Menten) for transdermal absorption.

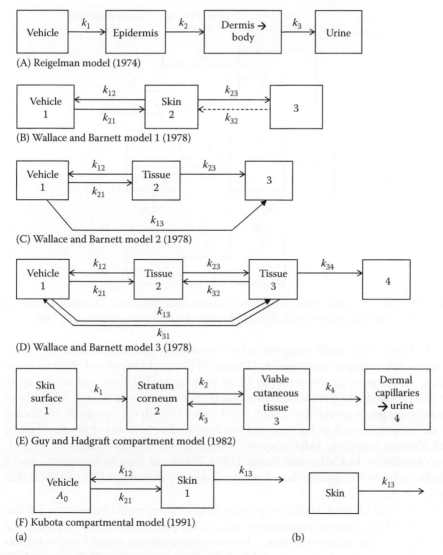

FIGURE 2.21 Selected compartment models in skin.

McCarley and Bunge (1998) modified this method and developed 11 one-compartment models. They matched various one-compartment model rate expressions with a diffusion model treating skin as a single membrane. Among the 11 models examined and matched, interestingly, the McCarley best predictor model claimed is almost identical to the Wallace and Barnett model 1 if the same assumptions are applied to both the models. Some observations were reported by McCarley after comparing these one-compartment models. For lipophilic chemicals, the ve provides significant resistance and longer time to reach steady states. The models using the membrane model characteristics often predict the membrane model

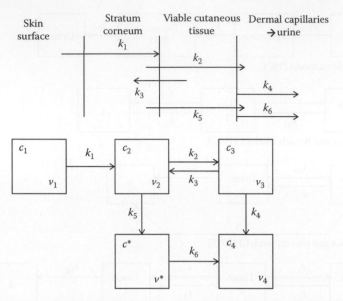

FIGURE 2.22 A compartment model for percutaneous absorption including epidermal metabolism. (Adapted from Guy, R.H. et al., *Toxicol. Appl. Pharmacol.*, 78, 123, 1985.)

results better. This work summarized systematically derived equations for coefficients of the compartment models in terms of physiochemical and physical properties, which were clearly arranged in an easy accessible form and are ready to be incorporated into a physiologically based pharmacokinetic model. A variety of two-compartment models have been developed to match the two-layer membrane model properties such as lag time and steady-state flux in terms of physiochemical and physical properties (McCarley and Bunge 2000). All one- and two-compartment models by McCarley and Bunge (1998, 2000) are easy to incorporate into a whole-body physiologically based pharmacokinetic model (PBPK), which is discussed in Section 2.3.

Kubota and Maibach (1992) proposed a compartment model rather than a diffusion model in finite-dose percutaneous absorption pharmacokinetics to predict the lag times and the steady-state flux. The intercompartmental transfer rate constants were also defined in their study and the rate constants were derived by matching the one-compartment and one-membrane model of the first and second statistical moments of the skin layer, but the model did not specify the layers included. Kubota et al. (1993) later utilized a simple single-layer model to estimate the permeability coefficient and the amount of drug in the skin at the steady state. Kubota and Maibach then developed a mathematical three-layer diffusion model, and longer lag time and half-life after vehicle removal in the epidermis and split-thickness skin were reported when compared to those in the sc (Kubota and Maibach 1994). Also, this model assumed that the vehicle is a well-stirred tank and single compartment parameters were used since it was assumed that the sc, ve, and de were combined to study the total amount of chemical in the skin, the total diffusion resistance, and the total lag time.

Two-compartment models are used when the permeability coefficient of the sc and ve and the storage ability of the skin cannot be neglected. In the literature, besides the Guy model (1982), the Williams model (Williams and Riviere 1995) showed the dramatic decline of chemical mass transfer from the vehicle by diffusion into the skin or evaporation. Shatkin and Chinery reported similar two-compartment models with all the rate constants in terms of the physicochemical and physical properties of skin with the same assumptions (Shatkin and Brown 1991; Chinery and Gleason 1993). Their models assume that the penetration rate into the sc is proportional to the steady-state sc permeability coefficient and that the solubility of the chemical in the blood and cutaneous blood flow rate determine the clearance of the chemicals from the skin. The only difference in these two models is one of the rate constant expressions. Shatkin estimated the permeability coefficient, while Chinery used specific values for the permeability coefficient. For lipophilic compounds, most of the compartment models overestimate two-membrane models for the mass transfer across the epi to the blood, except for the Shatkin model, because the membrane models predict a larger steady-state flux than the Shatkin model, which results in an underestimation of the mass of chemical across the epi.

2.3 PHYSIOLOGICALLY BASED PHARMACOKINETIC MODEL IN DERMAL ABSORPTION

PBPK models utilize differential equations to describe the mass balance of chemicals as the chemicals undergo absorption, distribution, metabolism, and excretion in the body. PBPK modeling can be considered a unique route of pharmacokinetic modeling with a focus on chemical disposition in various tissues. A PBPK model treats the body in terms of well-stirred tanks connected by blood flow. The tanks are equivalent to lumped series of organs that have similar blood flow as well as similar affinity to the chemicals. A typical PBPK model is shown in Figure 2.23. PBPK models are mathematical representations of biological processes that have been used in risk assessment by the chemical industry, environmental protection agents, and the pharmaceutical industry in drug discovery candidate selection. Since PBPK models are based on physiological and biochemical properties of the body and chemicals, they provide the opportunity for dose surface area, duration, and species extrapolation (e.g., from animal to human) in drug discovery and development. PBPK modeling received attention in 1937 when Teorell first adapted the approach to whole-body physiologically based modeling (Teorell 1937a,b). However, it was not until the 1960s, with the assistance of the computer, that PBPK modeling became active again and the representative models by Bellman et al. (1960), Bischoff and Brown (1966), and Dedrick and Bischoff (1968) first came about.

There are three anatomical regions of an individual organ: vascular, interstitial, and intracellular spaces. An individual organ can often be described either by a blood-flow-limited or by a membrane-limited model organ structure. In a PBPK model, depending on how the organ is "lumped," the organ structure can vary from a one-compartment to a three-subcompartment structure (Figure 2.24). It is assumed

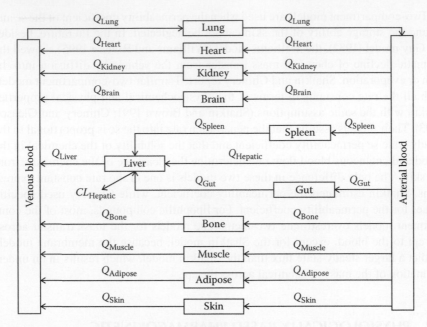

FIGURE 2.23 Schematic of a typical PBPK model.

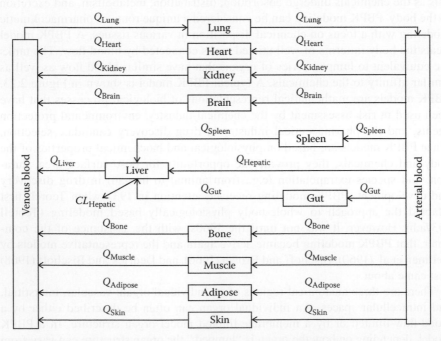

FIGURE 2.24 Subcompartments for organ structure in a PBPK model.

that either the subcompartment or the whole-organ compartment is a well-stirred homogeneous space with the chemical concentration being the same throughout the compartment.

2.3.1 MATHEMATICAL FORMULATIONS

The mathematical formulation in PBPK models have been summarized by Gallo (2002).

The differential mass balance equations for a noneliminating tissue compartment with three subcompartments are

$$V_i^V \frac{d}{dt} C_i^V = Q_i \left(C_p - C_i^V \right) - n_i^{VS}$$

$$V_i^S \frac{d}{dt} C_i^S = n_i^{VS} - n_i^{SI}$$

$$V_i^I \frac{d}{dt} C_i^I = n_i^{SI}$$

where

Q_i is the blood flow entering tissue i

C_i^V is the concentration in vascular space of tissue i

C_i^I is the concentration in interstitial space of tissue i

C_i^C is the concentration in intracellular space of tissue i

C_p is the blood concentration entering the tissue i

n_i^{VS} is the net mass flux from vascular space to interstitial space

n_i^{SI} is the net mass flux from interstitial space to intracellular space

V_i^V is the volume of vascular space in tissue i

V_i^S is the volume of interstitial space in tissue i

V_i^I is the volume of intracellular space in tissue i

2.3.1.1 Blood-Flow-Limited Model

For a blood-flow-limited model, vascular, interstitial, and intracellular spaces are lumped into a single homogeneous space

$$V_i \frac{d}{dt} C_i - Q_i \left(C_b \quad \frac{C_i}{R_i} \right)$$

where $R_i = C_i/C_0$, with C_0 as the plasma concentration at equilibrium.

2.3.1.2 Membrane-Limited Model

The three-subcompartment organ model can be simplified by lumping the vascular and interstitial subcompartments, and defined as an extracellular compartment.

Differential mass balance equations for a noneliminating membrane-limited compartment are

$$V_i^E \frac{d}{dt} C_i^E = Q_i \left(C_p - C_i^E \right) - n_i^{EI}$$

$$V_i^I \frac{d}{dt} C_i^I = n_i^{EI}$$

where subscript E indicates the extracellular space including both vascular and interstitial spaces.

2.3.1.3   Transport Mechanism

The transport of chemicals across biological membranes can be passive diffusion, carrier-mediated transport, or both. For passive diffusion, the flux across the membrane is described by a mass transfer coefficient (h_i) multiplying a concentration driving force:

$$n_i^{EI} = h_i \left(C_i^E - \frac{C_i^I}{R_i} \right)$$

$$n_i^{EI} = \left(\frac{a_i C_i^E}{b_i + C_i^E} \right) - \frac{\left(\dfrac{a_i C_i^I}{R_i} \right)}{\left(b_i + \dfrac{C_i^I}{R_i} \right)}$$

The total tissue compartment concentration C_i of a three-subcompartment organ structure can be expressed as

$$C_i = \frac{C_i^V V_i^V + C_i^S V_i^S + C_i^I V_i^I}{V_i}$$

The total tissue compartment concentration C_i in a two-subcompartment structure is

$$C_i = \frac{C_i^E V_i^E + C_i^I V_i^I}{V_i}$$

2.3.1.4   Excretion

For organs such as the liver and kidneys, excretion must be included in the following equation:

$$q_i = k_i V_i C_i$$

If the excretion kinetics are assumed to be saturable, then

$$q_i = \frac{V_m C_i}{K_m + C_i}$$

where

k_i is the first-order elimination rate constant (time^{-1})
V_m is the maximum elimination rate
K_m is the Michaelis–Menten constant

2.3.1.5 Metabolism

Oftentimes, the metabolism reaction will follow first-order or Michaelis–Menten saturable kinetics, so for a flow-limited compartment 1

$$V_i \frac{d}{dt} C_i = Q_i \left(C_b - \frac{C_i}{R_i} \right) - k_i C_i$$

$$V_i \frac{d}{dt} C_i = Q_i \left(C_b - \frac{C_i}{R_i} \right) - \left(\frac{V_m C_i}{K_m + C_i} \right)$$

2.3.2 Physiologically Based Pharmacokinetic Model for Dermal Absorption

There are often three types of dermal absorption models that can be incorporated into PBPK models: membrane models and one- and two-compartment models. Based on an earlier summary by Reddy (2005), the studies of dermal absorption in a PBPK model are summarized in Table 2.1.

Based on the surface area exposed, type of exposure, and the skin model chosen, the method to incorporate the dermal absorption model into the PBPK model can be selected. When only a small amount of skin is exposed to a chemical, the dermal absorption is often described as adding to the mixed venous blood, and skin is lumped as part of the slowly perfused compartment, such as for a study done for benzoic acid (Macpherson et al. 1996) and 2,3,7,8-tetrabromodibenzo-p-dioxin (Kedderis et al. 1993). If the whole body is exposed to a chemical, the skin is usually treated as a separate compartment for percutaneous absorption, such as a study done for chloroform exposure while swimming (Levesque et al. 2000) and whole-body exposures to m-xylene vapor in human (Loizou et al. 1999). Oftentimes, when PBPK models including a skin compartment are used for risk assessment, such as chloroform exposure during showering (McKone 1993), both dermal and inhalation exposure routes are included in the PBPK model. van der Merwe et al. (2006) used a PBPK skin model without incorporation into a whole-body PBPK model for organophosphate pesticides to describe in vitro dermal absorption through cells, and parameters related to the stratum corneum and solvent evaporation rates were estimated. Norman et al. (2008) employed and

TABLE 2.1
Physiologically Based Pharmacokinetic Models for Dermal Absorption

Chemical	Species	Model
Dichloromethane	Rat	McDougal et al. 1986
Bromochloromethane	Rat	McDougal et al. 1986
	Rat	Jepson and McDougal 1997
	Rat	Jepson and McDougal 1999
Dibromomethane	Rat	McDougal et al. 1986
	Rat	Bookout et al. 1996
	Rat	Bookout et al. 1997
	Rat	Jepson and McDougal 1997
	Rat	Jepson and McDougal 1999
Chloroform	Human	McKone 1993
	Human	Chinery and Gleason 1993
	Human	Georgopoulos et al. 1994
	Human	Roy et al. 1994
	Human	Roy et al. 1996a
	Human	Roy et al. 1996b
	Human	Corley et al. 2000
	Human	Levesque et al. 2000
	Human	Tan et al. 2006
	Human	Norman et al. 2008
Carbon tetrachloride	Rat	Thrall and Kenny 1996
Isopropanal	Rat	Clewell et al. 2001
Chlordecone	Rat	Heatherington et al. 1998
Chlorpyrifos	Human	Timchalk et al. 2002
Fluazifop-butyl	Human	Auton et al. 1994
Isofenphos	Rat	Knaak et al. 1994
Malathion	Human	Rabovsky and Brown 1993
	Human	Dong et al. 1994
2-Butoxyethanol	Rat	Shyr et al. 1993
	Human and rat	Corley et al. 1994
2-Butoxyethanol and butoxyacetic acid	Human	Franks et al. 2006
m-Xylene	Rat	McDougal et al. 1990
	Human	Loizou et al. 1999
o-Xylene	Human and rat	Thrall and Woodstock 2003
2,3,7,8-Tetrabromodibenzo-p-dioxin	Rat	Kedderis et al. 1993
Methyl t-butyl ether	Human and rat	Rao and Ginsberg 1997
Hexane	Rat	McDougal et al. 1990
Benzene	Rat	McDougal et al. 1990
	Human	Roy et al. 1998
Toluene	Rat	McDougal et al. 1990
	Rat	Thrall and Woodstock 2002
	Human	Thrall et al. 2002
Styrene	Rat	McDougal et al. 1990

TABLE 2.1 (continued)

Physiologically Based Pharmacokinetic Models for Dermal Absorption

Chemical	Species	Model
Isoflurane	Rat	McDougal et al. 1990
Halothane	Rat	McDougal et al. 1990
Perchoroethylene	Rat	McDougal et al. 1990
	Human	Rao and Brown 1993
	Human and rat	Poet et al. 2002
Trichoroethylene	Human and rat	Poet et al. 2000a
	Human	Haddad et al. 2006
Trihalomethanes	Human	Haddad et al. 2006
Methyl chloroform	Human	Bogen and Hall 1989
	Human and rat	Poet et al. 2000b
Octamethylcyclotetrasiloxane	Rat	Sarangapani et al. 2003
	Human	Reddy et al. 2007
Decamethylcyclopentasiloxane	Human	Reddy et al. 2007
Benzoic acid	Guinea pig	Macpherson et al. 1996
Ethylene glycol	Human and rat	Corley et al. 2005
Glycolic acid	Human and rat	Corley et al. 2005
Deltamethrin	Rat	Kim et al. 2008
Naphthalene	Human	Kim et al. 2007a
Organophosphate	Yorkshire pig	van der Merwe et al. 2006

compared three skin models: a skin compartment model (continuously stirred tank reactor model), an approximate membrane model (simplified time lag model), and a true membrane model (finite-difference model). All three models were incorporated in a PBPK model to predict chloroform in the breath during and after immersion in an aqueous solution (Norman et al. 2008). The study concluded that a well-stirred homogenous compartment skin model tends to predict more rapid initial absorption and lower cumulative absorption than does a membrane skin model. Norman et al. also concluded that permeability coefficient estimation depends on the skin model as well as the temperature of the aqueous solution. Corley et al. constructed a PBPK model that includes inhalation, oral, dermal, IV, and subcutaneous routes of administration to integrate the extensive mode of action of ethylene glycol and its intermediate metabolite glycolic acid (Corley et al. 2005). Franks et al. (2006) compared the simulation of butoxyacetic acid, the major metabolite of 2-butoxyethanol, in the blood during and after whole-body and dermal-only exposure.

2.3.2.1 Dermal Absorption Membrane Models

Percutaneous absorption through the skin is usually considered as a passive diffusion through one or two membranes (Scheuplein 1978). For hydrophilic chemicals that do not metabolize in the ve, dermal absorption can be represented by a one-layer membrane resistance of the sc. For highly lipophilic chemicals, both sc and ve should be considered as separate membranes (Cleek and Bunge 1993). Membrane models utilize Fick's law to describe the mass balance across the skin barrier. Even though membrane diffusion models are physiologically realistic, they can be much more mathematically complicated than well-stirred compartment models. Membrane models adapt partial differential equations that require numerical techniques and much more data resource, while compartment models use ordinary differential equations. Auton et al. considered skin as a two-layer membrane including both the sc and ve and incorporated it into a simple two-compartment systemic pharmacokinetic model (Auton et al. 1994).

Many studies (Georgopoulos et al. 1994; Roy et al. 1994, 1996a) have included a membrane model combined with a PBPK model to investigate human dermal exposures to chloroform. Roy et al. (1996b) compared three skin models that were incorporated into the same chloroform PBPK model. Roy and Georgopoulos (1998) reconstructed long-term exposures to benzene using PBPK model, and exhaled breath concentration was chosen as a biomarker of exposure. The study compared simulations of a one-membrane skin model and two-compartment skin models.

2.3.2.2   Dermal Absorption Compartment Models

Compartment models have been commonly used in the percutaneous absorption of chemicals. The most widely used model is shown in Figure 2.25. The mass balance transfer of chemicals across the skin from the vehicle to the blood depends on four rate constants (k_1, k_{-1}, k_2, k_{-2}, m³/s) and the concentration of chemicals in the vehicle (C_v), blood (C_b), and skin (C_{skin}). Physiologically relevant compartmental skin models, with rate constants all defined in terms of skin physiological properties such as partition coefficients and permeability coefficients, are readily to be combined with a whole-body PBPK model. The implementation of compartmental skin models with the first-order ordinary differential equations is commonly used for PBPK modeling. McCarley and Bunge (2001) reviewed 9 one-compartment and 2 two-compartment models for dermal absorption, with all of the coefficients expressed in terms of physiochemical and physical properties, which can be used in whole-body PBPK models.

A membrane model has been incorporated into a physiologically based toxicokinetic model that consists of five compartments: sc, ve, blood, fat, and other tissues, to demonstrate dermal and inhalation exposure of naphthalene (Kim et al. 2007a). A similar method of dermal absorption as a one-directional diffusive process according to Fick's first law of diffusion has been used a more complex whole-body PBPK model to characterize tertiary-butyl ether exposure (Kim et al. 2007b).

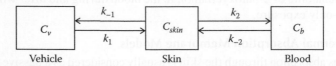

FIGURE 2.25   A schematic diagram of one-compartment model for skin used by McDougal et al. (1986) in a PBPK model. (Adapted from McDougal, J.N. et al., *Toxicol. Appl. Pharmacol.*, 85, 286, 1986.)

2.3.2.2.1 Dermal Absorption One-Compartment Models

The first model that incorporated dermal absorption into a whole-body PBPK model was a one-compartment model of all skin layers lumped together as a homogenous compartment by McDougal et al. (1986) (Figure 2.26). In this model, the skin compartment included a description of Fick's law to provide an input function (McDougal et al. 1986). The equation for a well-stirred skin compartment was shown as

$$V_{sk}\frac{dC_{sk}}{d_t} = Q_{sk}\left(C_b - \frac{C_{sk}}{R_{sk/b}}\right) + P_{sk}A_{sk}\left(C_{sfc} - \frac{C_{sk}}{R_{sk/sfc}}\right)$$

where

the subscript sk stands for skin and sfc stands for surface
P_{sk} is the permeability coefficient of the skin
A_{sk} is the skin surface area exposed

The first term on the right-hand side of the equation explains the amount of chemical in the skin contributed by the blood flow, while the second term explains the amount of chemical in the skin contributed by diffusion.

In this study, McDougal et al. used Fischer 344 rats to investigate dichloromethane (DCM), bromochloromethane (BCM), and dibromomethane (DBM) dermal absorption from the vapor phase. In a later study, they used the same skin model in a PBPK model for styrene, toluene, benzene, hexane, isoflurane, perchloroethylene, *m*-xylene, and halothane dermal absorption from the vapor phase (McDougal et al. 1990). Later, similar skin models have been used to study the dermal absorption of DBM and BCM from aqueous solutions, water, mineral oil, and corn oil vehicles (Jepson and McDougal 1997, 1999).

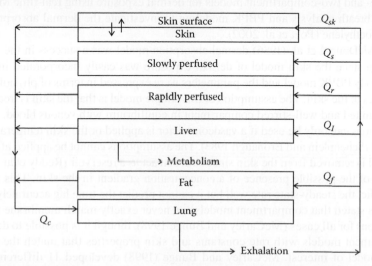

FIGURE 2.26 Schematic diagram of the PBPK model used by McDougal et al. (1986). (Adapted from McDougal, J.N. et al., *Toxicol. Appl. Pharmacol.*, 85, 286, 1986.)

Bookout et al. (1996) studied DBM with skin model as a homogenous, well-stirred compartment with two subcompartments: sc as one subcompartment and a subcompartment that includes ve, de, and subcutaneous fat. In a later similar study, a shunt pathway of DBM through appendages was included in the second skin subcompartment (Bookout et al. 1996, 1997).

Several other studies have employed the McDougal et al. (1986) dermal absorption model incorporated into a whole-body PBPK model. Corley adapted this skin model as dermal absorption into a PBPK model with IV infusion, inhalation, oral, and dermal exposure routes for 2-butoxyethanol in human and rats (Corley et al. 1994). Corley later used a similar approach to describe the chloroform exposure with one more route via intraperitoneal injection. Rao and Brown used the McDougal et al. (1986) skin model for dermal exposure in a PBPK model of tetrachloroethlyene in both aqueous and gas phases (Rao and Brown 1993). In a later study (Rao and Ginsberg 1997), the dermal absorption of *t*-butyl ether and its metabolite *t*-butyl alcohol was determined the same way. The skin model of McDougal et al. (1986) has been also used for assessment of water contaminants: trichloroethylene and trihalomethanes (Haddad et al. 2006). In a study by Tan et al. (2006), the same model was used as in the study by Corley et al. (2000) for human exposure to chloroform. In the last decade, real-time analysis has been combined with PBPK modeling. Real-time breath concentration data can be obtained by measurement using an ion-trap mass spectrometer. Thrall et al. (2002) used PBPK models to estimate the permeability coefficients of methyl chloroform, trichloroethylene, benzene, and toluene in rats, monkeys, and humans by analyzing exhaled breath concentration after dermal exposure using a real-time MS–MS technique and the skin model by McDougal et al. (1986). The same real-time method has been used to investigate the percutaneous absorption of trichloroethylene and permeability coefficients estimation of methyl chloroform absorption across skin from soil and water in rats and humans (Poet et al. 2000a,b). Poet et al. later employed both one- and two-compartment models for dermal exposure using real-time MS/MS exhaled breath analysis and PBPK modeling to investigate the dermal absorption of perchloroethylene (Poet et al. 2002).

The McDougal et al. (1986) dermal absorption model was a success in the PBPK modeling since the skin model of dermal exposure was easily incorporated into the whole-body PBPK model and the parameters were expressed in terms of physiological properties of the skin. The assumption of this simple model is that the skin is treated as a flow-limited and well-stirred compartment in equilibrium with venous blood. These assumptions can only be used if a vasoconstrictor is applied or the skin temperature is declined (Scheuplein and Bronaugh 1983). The assumptions cannot be applied after the chemical is removed from the skin since skin can act as a reservoir (Reddy et al. 1998) because of the possible presence of a concentration gradient in the skin. This model can predict the steady-state rate well but it cannot predict the time lag accurately.

It was stated that compartment models can never exactly match membrane model predictions for all cases (McCarley and Bunge 1998), though it is possible to develop compartment models with rate constants and skin properties that match the membrane model of interest. McCarley and Bunge (1998) developed 11 different one-compartment skin models that are physiologically relevant to match the different membrane model cases.

2.3.2.2.2 Dermal Absorption Two-Compartment Models

Two-compartment models have been incorporated into a PBPK model as well. Chinery and Gleason (1993) used a two-compartment skin model for the dermal absorption in a PBPK model to study chloroform exposure. The two compartments in this model represent sc and ve, respectively, but the ve did not provide a rate-limiting resistance for chloroform transport, which is the reason most of the chloroform absorption has been reported using one-compartment skin model. One-compartment model should be employed when hydrophilic and low lipophilic compounds are studied since the ve does not provide rate-limiting resistance to percutaneous absorption. McCarley and Bunge (2000) have developed various two-compartment models to match two-layer membrane model properties such as lag time and steady-state flux.

Heatherington incorporated the (Guy et al. 1985) two-compartment model into a whole-body PBPK model for lipophilic solid-form chlordecone distribution (Heatherington et al. 1998). The solid chlordecone formed on the skin surface because of the evaporation of acetone. Therefore, in this case, the solubility parameter should be used instead of the partition coefficient.

A two-compartment skin model (skin and deep skin compartments) were incorporated into an inhalation PBPK model by Reddy et al. (2007) to interpret human dermal absorption of octamethylcyclotetrasiloxane (D4) and decamethylcyclopentasiloxane (D5) through axilla. The two-compartmental model included volatilization of chemicals from the skin surface, diffusion of absorbed chemicals back to the skin surface, evaporation of chemicals from the skin surface after the removal of the chemicals, and the transfer of chemicals into the blood and storage compartment within the skin.

2.4 CUTANEOUS MICRODIALYSIS

There are a few techniques that are available for evaluating topical drug absorption in the skin. Lately, several studies reported dermatophamacokinetics studies using tape-stripping technique to access the topical bioavailability of drugs in human skin (N'Dri-Stempfer et al. 2009; Nicoli et al. 2009; Wiedersberg et al. 2009a,b). Other traditional methods such as punch biopsies, shave biopsies, and suction blisters can be used to assess cutaneous drug concentration but are not feasible in humans due to their invasive nature or difficulty to conduct.

Cutaneous microdialysis (MD) allows continuous real-time sampling in the skin and monitoring the drug concentration in the tissue. MD is conducted under minimally invasive ways when compared with traditional and invasive methods above, and is becoming an established valuable technique for pharmacokinetics and bioequivalence studies in human skin (Benfeldt et al. 2007; Ortiz et al. 2008; Schmidt et al. 2008). The MD probe consists of a semipermeable membrane and is constantly perfused with a physiological solution at low flow rates of approximately 0.1–5 μL/min (Chaurasia et al. 2007). Free drug molecules can pass through the semipermeable membrane by passive diffusion driven by the concentration gradients (de Lange et al. 1997; Chaurasia et al. 2007). The amount of drug removed by the perfused solution is correlated with the total amount absorbed.

Ortiz et al. (2008) recently reported the first study of transdermal drug penetration of a metronidazole cream in atopic dermatitis skin using MD and tape-stripping techniques. In this study, both methods were able to sample topical drug concentration in the disease skin, and found that a 2.4-fold increase in dermal metronidazole concentration caused by sparse atopic dermatitis was detected by MD sampling, but not by tape stripping. However, the presence of the disease resulted in double interindividual variability with cutaneous MD sampling in skin and no significant effect on tape stripping variability. The study concluded that cutaneous MD seems to be the most feasible method for topical drug penetration in atopic dermatitis.

A more detailed discussion on the application of cutaneous MD in the study of dermatokinetics of drugs is provided in Chapter 5.

REFERENCES

Albery, W. J. and Hadgraft, J. 1979. Percutaneous absorption: Theoretical description. *J Pharm Pharmacol* 31(3): 129–139.

Auton, T. R., D. R. Westhead et al. 1994. A physiologically based mathematical model of dermal absorption in man. *Hum Exp Toxicol* 13(1): 51–60.

Bellman, R., J. A. Jacquez et al. 1960. Some mathematical aspects of chemotherapy. I: One organ models. *Bull Math Biophys* 22: 181–198.

Benfeldt, E., S. H. Hansen et al. 2007. Bioequivalence of topical formulations in humans: Evaluation by dermal microdialysis sampling and the dermatopharmacokinetic method. *J Invest Dermatol* 127(1): 170–178.

Bischoff, K. B. and Brown, R. G. 1966. Drug distribution in mammals. *Chem Eng Prog Symp* 62: 33–45.

Bogen, K. T. and Hall, L. C. 1989. Pharmacokinetics for regulatory risk analysis: The case of 1,1,1-trichloroethane (methyl chloroform). *Regul Toxicol Pharmacol.* 10(1): 26-50.

Bookout, R. L., Jr., C. R. McDaniel et al. 1996. Multilayered dermal subcompartments for modeling chemical absorption. *SAR QSAR Environ Res* 5(3): 133–150.

Bookout, R. L., Jr., D. W. Quinn et al. 1997. Parallel dermal subcompartments for modeling chemical absorption. *SAR QSAR Environ Res* 7(1–4): 259–279.

Chaurasia, C. S., M. Muller et al. 2007. AAPS-FDA Workshop White Paper: Microdialysis principles, application, and regulatory perspectives. *J Clin Pharmacol* 47(5): 589–603.

Chinery, R. L. and A. K. Gleason 1993. A compartmental model for the prediction of breath concentration and absorbed dose of chloroform after exposure while showering. *Risk Anal* 13(1): 51–62.

Cleek, R. L. and A. L. Bunge 1993. A new method for estimating dermal absorption from chemical exposure. 1. General approach. *Pharm Res* 10(4): 497–506.

Clewell, H. J., 3rd, Gentry, P. R., Gearhart, J. M., Covington, T. R., Banton, M. I., and Andersen, M. E. 2001. Development of a physiologically based pharmacokinetic model of isopropanol and its metabolite acetone. *Toxicol Sci.* 63(2): 160–172.

Corley, R. A., M. J. Bartels et al. 2005. Development of a physiologically based pharmacokinetic model for ethylene glycol and its metabolite, glycolic acid, in rats and humans. *Toxicol Sci* 85(1): 476–490.

Corley, R. A., G. A. Bormett et al. 1994. Physiologically based pharmacokinetics of 2-butoxyethanol and its major metabolite, 2-butoxyacetic acid, in rats and humans. *Toxicol Appl Pharmacol* 129(1): 61–79.

Corley, R. A., S. M. Gordon et al. 2000. Physiologically based pharmacokinetic modeling of the temperature-dependent dermal absorption of chloroform by humans following bath water exposures. *Toxicol Sci* 53(1): 13–23.

de Lange, E. C., M. Danhof et al. 1997. Methodological considerations of intracerebral micro-dialysis in pharmacokinetic studies on drug transport across the blood-brain barrier. *Brain Res Brain Res Rev* 25(1): 27–49.

Dedrick, R. L. and Bischoff, K. B. 1968. Pharmacokinetics in applications of the artificial kidney. *Chem. Eng. Prog. Symp. Ser.* 64: 32–44.

Derendorf, H., T. Gramatté et al. 2002. *Pharmakokinetik.* eds: Derendorf, H., Gramatté, T., Schäfer, G. Stuttgart, Germany: Wissenschaftliche Verlagsgesellschaft mbH.

Dong, M. H., W. M. Draper et al. 1994. Estimating malathion doses in California's medfly eradication campaign using a physiologically based pharmacokinetic model. *Adv Chem Ser* 241: 189–208.

el Tayar, N., R. S. Tsai et al. 1991. Percutaneous penetration of drugs: A quantitative structure-permeability relationship study. *J Pharm Sci* 80(8): 744–749.

Franks, S. J., M. K. Spendiff et al. 2006. Physiologically based pharmacokinetic modelling of human exposure to 2-butoxyethanol. *Toxicol Lett* 162(2–3): 164–173.

Gallo, J. M. 2002. Physiologically based pharmacokinetic models. In: *Pharmacokinetics in Drug Discovery and Development.* ed.: Ronald D. Schoenwald. Boca Raton, FL: CRC Press.

Georgopoulos, P. G., A. Roy et al. (1994). Reconstruction of short-term multi-route exposure to volatile organic compounds using physiologically based pharmacokinetic models. *J Expo Anal Environ Epidemiol* 4: 309–328.

Guy, R. H. and Hadgraft, J. 1982. Percutaneous metabolism with saturable enzyme kinetics. *Int J Pharm* 11: 187–197.

Guy, R. H. and Hadgraft, J. 1984. Pharmacokinetics of percutaneous absorption with concurrent metabolism. *Int J Pharm* 20: 43–51.

Guy, R. H., J. Hadgraft et al. 1982. A pharmacokinetic model for percutaneous absorption. *Int J Pharm* 11: 119–129.

Guy, R. H., J. Hadgraft et al. 1983. Percutaneous absorption: Multidose pharmacokinetics. *Int J Pharm* 17: 23–28.

Guy, R. H., J. Hadgraft et al. 1985. Percutaneous absorption in man: A kinetic approach. *Toxicol Appl Pharmacol* 78(1): 123–129.

Haddad, S., G. C. Tardif et al. 2006. Development of physiologically based toxicokinetic models for improving the human indoor exposure assessment to water contaminants: Trichloroethylene and trihalomethanes. *J Toxicol Environ Health A* 69(23): 2095–2136.

Heatherington, A. C., H. L. Fisher et al. 1998. Percutaneous absorption and disposition of [14C]chlordecone in young and adult female rats. *Environ Res* 79(2): 138–155.

Hostynek, J. J., W. G. Reifenrath et al. 1995. Correlation of in vivo and in vitro percutaneous absorption with a mathematical model. *Curr Probl Dermatol* 22: 139–145.

Jepson, G. W. and J. N. McDougal. 1997. Physiologically based modeling of nonsteady state dermal absorption of halogenated methanes from an aqueous solution. *Toxicol Appl Pharmacol* 144(2): 315–324.

Jepson, G. W. and J. N. McDougal. 1999. Predicting vehicle effects on the dermal absorption of halogenated methanes using physiologically based modeling. *Toxicol Sci* 48(2): 180–188.

Kalia, Y. N. and R. H. Guy. 2001. Modeling transdermal drug release. *Adv Drug Deliv Rev* 48(23): 159–172.

Kedderis, L. B., J. J. Mills et al. 1993. A physiologically based pharmacokinetic model for 2,3,7,8-tetrabromodibenzo-p-dioxin (TBDD) in the rat: Tissue distribution and CYP1A induction. *Toxicol Appl Pharmacol* 121(1): 87–98.

Kim, D., M. E. Andersen et al. 2007a. PBTK modeling demonstrates contribution of dermal and inhalation exposure components to end-exhaled breath concentrations of naphthalene. *Environ Health Perspect* 115(6): 894–901.

Kim, D., M. E. Andersen et al. 2007b. Refined PBPK model of aggregate exposure to methyl tertiary-butyl ether. *Toxicol Lett* 169(3): 222–235.

Kim, K. B., S. S. Anand et al. 2008. Toxicokinetics and tissue distribution of deltamethrin in adult Sprague-Dawley rats. *Toxicol Sci* 101(2): 197–205.

Knaak, J. B., M. A. Al-Bayati et al. 1994. Prediction of anticholinesterase activity and urinary metabolites of isofenphos: Use of apercutaneous physiologically based pharmacokinetic-physiologically based pharmacodynamic model. In: *Biomarkers of Human Exposure to Pesticides*, Chapter 17. eds: Saleh, M. A., Blancato, J. N., Nauman, C. H. *ACS Symp Ser* 542, Washington, DC.

Kubota, K. and H. I. Maibach. 1991. A compartment model for percutaneous drug absorption. *J Pharm Sci* 80(5): 502–504.

Kubota, K. and H. I. Maibach. 1992. A compartment model for percutaneous absorption: Compatibility of lag time and steady-state flux with diffusion model. *J Pharm Sci* 81(9): 863–865.

Kubota, K. and H. I. Maibach. 1994. Significance of viable skin layers in percutaneous permeation and its implication in mathematical models: Theoretical consideration based on parameters for betamethasone 17-valerate. *J Pharm Sci* 83(9): 1300–1306.

Kubota, K., J. Ademola et al. 1995. Simultaneous diffusion and metabolism of betamethasone 17-valerate in the living skin equivalent. *J Pharm Sci* 84(12): 1478–1481.

Kubota, K., M. Sznitowska et al. 1993. Percutaneous absorption: A single-layer model. *J Pharm Sci* 82(5): 450–456.

Levesque, B., P. Ayotte et al. 2000. Evaluation of the health risk associated with exposure to chloroform in indoor swimming pools. *J Toxicol Environ Health A* 61(4): 225–243.

Loizou, G. D., K. Jones et al. 1999. Estimation of the dermal absorption of m-xylene vapor in humans using breath sampling and physiologically based pharmacokinetic analysis. *Toxicol Sci* 48(2): 170–179.

Macpherson, S. E., C. N. Barton et al. 1996. Use of in vitro skin penetration data and a physiologically based model to predict in vivo blood levels of benzoic acid. *Toxicol Appl Pharmacol* 140(2): 436–443.

McCarley, K. D. and A. L. Bunge. 1998. Physiologically relevant one-compartment pharmacokinetic models for skin. 1. Development Of models. *J Pharm Sci* 87(10): 1264.

McCarley, K. D. and A. L. Bunge. 2000. Physiologically relevant two-compartment pharmacokinetic models for skin. *J Pharm Sci* 89(9): 1212–1235.

McCarley, K. D. and A. L. Bunge. 2001. Pharmacokinetic models of dermal absorption. *J Pharm Sci* 90(11): 1699–1719.

McDougal, J. N., G. W. Jepson et al. 1986. A physiological pharmacokinetic model for dermal absorption of vapors in the rat. *Toxicol Appl Pharmacol* 85(2): 286–294.

McDougal, J. N., G. W. Jepson et al. 1990. Dermal absorption of organic chemical vapors in rats and humans. *Fundam Appl Toxicol* 14(2): 299–308.

McKone, T. E. 1993. Linking a PBPK model for chloroform with measured breath concentrations in showers: Implications for dermal exposure models. *J Expo Anal Environ Epidemiol* 3(3): 339–365.

Michaels, A. S., Chandrasekaran, S. K., and Shaw, S. E. 1975. Drug permeation through human skin: Theory and in vitro experimental measurement. *AIChE Journal* 21(5): 985–996.

N'Dri-Stempfer, B., W. C. Navidi et al. 2009. Improved bioequivalence assessment of topical dermatological drug products using dermatopharmacokinetics. *Pharm Res* 26(2): 316–328.

Nicoli, S., A. L. Bunge et al. 2009. Dermatopharmacokinetics: Factors influencing drug clearance from the stratum corneum. *Pharm Res* 26(4): 865–871.

Norman, A. M., J. C. Kissel et al. 2008. Effect of PBPK model structure on interpretation of in vivo human aqueous dermal exposure trials. *Toxicol Sci* 104(1): 210–217.

Ortiz, P. G., S. H. Hansen et al. 2008. The effect of irritant dermatitis on cutaneous bioavailability of a metronidazole formulation, investigated by microdialysis and dermatopharmacokinetic method. *Contact Dermatitis* 59(1): 23–30.

Poet, T. S., R. A. Corley et al. 2000a. Assessment of the percutaneous absorption of trichloroethylene in rats and humans using MS/MS real-time breath analysis and physiologically based pharmacokinetic modeling. *Toxicol Sci* 56(1): 61–72.

Poet, T. S., K. D. Thrall et al. 2000b. Utility of real time breath analysis and physiologically based pharmacokinetic modeling to determine the percutaneous absorption of methyl chloroform in rats and humans. *Toxicol Sci* 54(1): 42–51.

Poet, T. S., K. K. Weitz et al. 2002. PBPK modeling of the percutaneous absorption of perchloroethylene from a soil matrix in rats and humans. *Toxicol Sci* 67(1): 17–31.

Rabovsky, J. and Brown, J. P. 1993. Malathion metabolism and disposition in mammals. *J. Occup. Med. Toxicol.* 2, 131–168.

Rao, H. V. and D. R. Brown. 1993. A physiologically based pharmacokinetic assessment of tetrachloroethylene in groundwater for a bathing and showering determination. *Risk Anal* 13(1): 37–49.

Rao, H. V. and G. L. Ginsberg. 1997. A physiologically-based pharmacokinetic model assessment of methyl t-butyl ether in groundwater for a bathing and showering determination. *Risk Anal* 17(5): 583–598.

Reddy, M. B. 2005. Dermal exposure models. In *Physiologically Based Pharmacokinetic Modeling: Science and Applications.* eds: Reddy, M. B., Yang, R. S. H., Clewell III, H. J., Andersen, M. E., pp. 375–385. Hoboken, NJ: John Wiley & Sons.

Reddy, M. B., R. J. Looney et al. 2007. Modeling of human dermal absorption of octamethylcyclotetrasiloxane (D(4)) and decamethylcyclopentasiloxane (D(5)). *Toxicol Sci* 99(2): 422–431.

Reddy, M. B., K. D. McCarley et al. 1998. Physiologically relevant one-compartment pharmacokinetic models for skin. 2. Comparison of models when combined with a systemic pharmacokinetic model. *J Pharm Sci* 87(4): 482–490.

Riegelman, S. 1974. Pharmacokinetics. Pharmacokinetic factors affecting epidermal penetration and percutaneous adsorption. *Clin Pharmacol Ther* 16(5 Part 2): 873–883.

Roberts, M. S. and Anissimov Y. G. Mathematical models in percutaneous absorption. In: Percutaneous Absorption Drug, Cosmetics, Mechanisms, Methods. Fourth edition. Ed: Bronaugh, R. L. and Maibach, H. I. New York, NY: Marcel Dekker, Inc.

Roy, A. and P. G. Georgopoulos et al. 1998. Reconstructing week-long exposures to volatile organic compounds using physiologically based pharmacokinetic models. *J Expo Anal Environ Epidemiol* 8(3): 407–422.

Roy, A., C. P. Weisel et al. 1994. Studies of multiroute exposure/dose reconstruction using physiologically based pharmacokinetic models. In: *Hazardous Wastes and Public Health.* ed.: Andrews, J. S., pp. 284–293. Atlanta, GA.

Roy, A., C. P. Weisel et al. 1996a. Studies of multiroute exposure/dose reconstruction using physiologically based pharmacokinetic models. *Toxicol Ind Health* 12(2): 153–163.

Roy, A., C. P. Weisel et al. 1996b. A distributed parameter physiologically-based pharmacokinetic model for dermal and inhalation exposure to volatile organic compounds. *Risk Anal* 16(2): 147–160.

Sarangapani, R., J. Teeguarden et al. 2003. Route-specific differences in distribution characteristics of octamethylcyclotetrasiloxane in rats: Analysis using PBPK models. *Toxicol Sci* 71(1): 41–52.

Scheuplein, R. J. 1978. Permeability of the skin: A review of major concepts. *Curr Probl Dermatol* 7: 172–186.

Scheuplein, R. J. and Bronaugh, R. L. 1983. Percutaneous absorption. In: *Biochemistry and Physiology of the Skin.* ed.: Goldsmith, L. A. New York: Oxford University Press, pp. 1255–1295.

Schmidt, S., R. Banks et al. 2008. Clinical microdialysis in skin and soft tissues: An update. *J Clin Pharmacol* 48(3): 351–364.

Shatkin, J. A. and H. S. Brown. 1991. Pharmacokinetics of the dermal route of exposure to volatile organic chemicals in water: A computer simulation model. *Environ Res* 56(1): 90–108.

Shyr, L. J., P. J. Sabourin et al. 1993. Physiologically based modeling of 2-butoxyethanol disposition in rats following different routes of exposure. *Environ Res* 63(2): 202–218.

Tan, Y. M., K. H. Liao et al. 2006. Use of a physiologically based pharmacokinetic model to identify exposures consistent with human biomonitoring data for chloroform. *J Toxicol Environ Health A* 69(18): 1727–1756.

Teorell, T. 1937a. Kinetics of distribution of substances administered to the body. I. The extravascular modes of administration. *Arch Int Pharmacodyn Ther* 57: 205–225.

Teorell, T. 1937b. Kinetics of distribution of substances administered to the body. II. The intravascular modes of administration. *Arch Int Pharmacodyn Ther* 57: 226–240.

Thral, K. D. and Kenny, D. V. 1996. Evaluation of a carbon tetrachloride physiologically based pharmacokinetic model using real-time breath-analysis monitoring of the rat. *Inhal. Toxicol.* 8: 251–261.

Thrall, K. D. and A. D. Woodstock. 2002. Evaluation of the dermal absorption of aqueous toluene in F344 rats using real-time breath analysis and physiologically based pharmacokinetic modeling. *J Toxicol Environ Health A* 65(24): 2087–2100.

Thrall, K. and A. Woodstock. 2003. Evaluation of the dermal bioavailability of aqueous xylene in F344 rats and human volunteers. *J Toxicol Environ Health A* 66(13): 1267–1281.

Thrall, K. D., K. K. Weitz et al. 2002. Use of real-time breath analysis and physiologically based pharmacokinetic modeling to evaluate dermal absorption of aqueous toluene in human volunteers. *Toxicol Sci* 68(2): 280–287.

Timchalk, C., R. J. Nolan et al. 2002. A Physiologically based pharmacokinetic and pharmacodynamic (PBPK/PD) model for the organophosphate insecticide chlorpyrifos in rats and humans. *Toxicol Sci* 66(1): 34–53.

van der Merwe, D., J. D. Brooks et al. 2006. A physiologically based pharmacokinetic model of organophosphate dermal absorption. *Toxicol Sci* 89(1): 188–204.

Wallace, S. M. and G. Barnett. 1978. Pharmacokinetic analysis of percutaneous absorption: Evidence of parallel penetration pathways for methotrexate. *J Pharmacokinet Biopharm* 6(4): 315–325.

Wiedersberg, S., C. S. Leopold et al. 2009a. Dermatopharmacokinetics of betamethasone 17-valerate: Influence of formulation viscosity and skin surface cleaning procedure. *Eur J Pharm Biopharm* 71(2): 362–366.

Wiedersberg, S., A. Naik et al. 2009b. Pharmacodynamics and dermatopharmacokinetics of betamethasone 17-valerate: Assessment of topical bioavailability. *Br J Dermatol* 160(3): 676–686.

Williams, P. L. and J. E. Riviere. 1995. A biophysically based dermatopharmacokinetic compartment model for quantifying percutaneous penetration and absorption of topically applied agents. I. Theory. *J Pharm Sci* 84(5): 599–608.

3 Formulation Approaches to Modulate the Dermatokinetics of Drugs

Srinivasa Murthy Sammeta, Michael A Repka, and S. Narasimha Murthy

CONTENTS

3.1 INTRODUCTION

Skin is the most accessible and largest organ in the body. There has been a long history of application of medicaments to the skin for local, regional, and systemic effects. In the case of formulations intended for local activity, the contents are expected to be retained on the skin and not penetrate into deeper layers. On the other hand, while treating the cutaneous and subcutaneous disorders, the medication is required to penetrate into the affected tissue. Medicated creams, ointments, lotions, gels, and sprays constitute the formulations intended for local and regional delivery of medicaments. Specially designed vehicles are also available to target the delivery of actives specifically to pilosebaceous structures, dermal–epidermal junction, and subdermal tissues. The drugs administered for the treatment of cutaneous and subcutaneous disorders include anti-infective agents, local anesthetics, nonsteroidal anti-inflammatory drugs, and steroids. Systemic delivery of drugs via the transdermal route has gained a lot of importance in the last two decades. A number of transdermal patch systems for the treatment of angina, motion sickness, hypertension, smoking cessation, and pain have been successful in the market.

The most challenging task in the delivery of drugs into skin is to overcome the strong permeation barrier, residing in the upper layers of the skin. Extremely hydrophilic or lipophilic molecules encounter resistance to their penetration into skin layers. Generally, extremely hydrophilic drugs are poorly permeable across the stratum corneum, whereas highly lipophilic drugs tend to be retained in the lipid domains of the stratum corneum. Therefore, a molecule with the log $P_{oct/water}$ of 1–2 is considered as the most suitable for delivery into skin. To enhance the rate of penetration of poorly permeable drugs, several physiological, physicochemical, and formulation approaches are sought.

3.2 FORMULATION APPROACHES TO MODULATE DERMATOKINETICS OF DRUGS

Some of the most investigated and implemented physicochemical approaches for enhancing drug delivery into skin are ion pairing, forming eutectic mixtures, supersaturation, and prodrug formation (Figure 3.1). The ionized drugs generally possess a low partition coefficient. Ion pairing is known to enhance the permeation by increasing the partitioning of drug into the lipid region of the stratum corneum (Barry 2001). On reaching the aqueous viable epidermis, the ion pair dissociates releasing the parent drug ion, which undergoes further diffusion into deeper layers (Megwa et al. 2000a,b; Stott et al. 2001; Valenta et al. 2000).

The melting point of the drugs is also related to the transdermal flux. This relationship was clearly deduced initially by Potts and Guy and later by others (Benson 2005; Calpena et al. 1994; Potts and Guy 1992). Reduction of melting point by way of preparing eutectic mixtures was thus found to enhance the drug

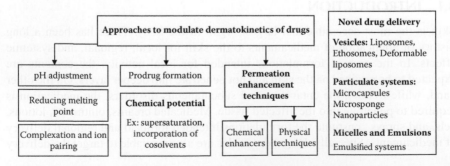

FIGURE 3.1   Various approaches to modulate dermatokinetics of drugs.

penetration into skin. The eutectic mixture of lignocaine and prilocaine (well known as EMLA cream) is known to provide effective anesthesia for pain-free venepuncture when applied under occlusive conditions (Benson 2005). Eutectic systems formed between ibuprofen and terpenes were found to enhance the delivery of ibuprofen (Stott et al. 1998).

Supersaturation of drug in the formulation is associated with a high thermodynamic activity; hence, this approach has been investigated as a potential way to enhance the cutaneous drug delivery. Different techniques have been used to produce supersaturated states such as solvent evaporation, mixing of two solvents (wherein the drug solubility in one is much more than in the other) or by mixing of two solutions of different pH (where the solubility is pH dependent). Megrab et al. reported an 18-fold increase in the stratum corneum uptake of estradiol with supersaturated solution over a saturated solution that was 18 times less concentrated (Megrab et al. 1995).

Prodrugs have been utilized to enhance the delivery of drugs with unfavorable partition coefficients (Sloan 1992). The prodrug approach involves the addition of a promoiety to increase the partition coefficient and hence permeation of the parent molecule into deeper skin layers. A prodrug, after diffusion into the skin, undergoes reversion to the active drug by the metabolic processes. For example, the permeability coefficient of the parent compound, indomethacin, has been reported to increase by 100-fold by conversion to an ester prodrug (Jona et al. 1995). The intrinsic permeability of 5-fluorouracil was found to improve by conversion to 1-alkyloxycarbonyl-5-fluorouracil prodrug (Beall and Sloan 1996). The ethyl derivative was found to be the most effective with a 25-fold increase in the flux across hairless mouse skin when compared to 5-fluorouracil (Beall et al. 1994).

The adjustment of pH to increase the unionized fraction of drug in the formulation, incorporation of a co-solvent to enhance the solubility of drugs in the formulation, and incorporation of permeation enhancers in the formulation have been the most widely used formulation approaches to enhance drug delivery into the skin. Various classes of permeation enhancers and their mechanism of interaction with the stratum corneum have been well documented in the literature so far (Barry 1988; Williams and Barry 2004). Most of the permeation enhancers are known to alter the organization of lipids or change the physiochemical properties of skin to enhance the drug penetration into skin. Some enhancers penetrate into skin to affect the metabolism of drugs. The azone, L-menthol, ethanol, and the menthol–ethanol–water systems have been found to decrease the metabolism of ethyl nicotinate in the skin (Hayashi et al. 1997).

3.3 NOVEL DRUG DELIVERY APPROACHES

Several novel drug delivery approaches have been investigated for effective cutaneous delivery of drugs. The most investigated were the carrier-mediated systems. The carrier-mediated systems have been demonstrated to provide modified drug release and targeting of drugs to specific structures in the skin. They were also found to protect some drugs from undergoing metabolism in the skin.

3.3.1 Vesicles as Drug Carriers

3.3.1.1 Liposomes

Liposomes are defined as vesicles of lipids that enclose an aqueous volume within. Liposomes are generally formulated using phospholipids. Depending on the size and structure, liposomes could be formulated as unilamellar or multilamellar vesicles ranging from nano to micron size (Figure 3.2) (Crommelin and Schreier 1994; Samad et al. 2007; Williams 2003).

Mezei and Gulasekharam first reported the application of liposomes as cutaneous drug delivery systems in 1980. They found that the liposomal lotion developed using dipalmitoyl-phosphatidylcholine enhanced the epidermal and dermal levels of triamcinolone acetonide by approximately four- to fivefold over the conventional dosage form (Mezei and Gulasekharam 1980, 1982). Foldvari and coworkers observed a greater local anesthetic effect in human volunteers when liposomal formulation of lidocaine was applied on the forearm compared to the cream formulation due to improved accumulation of the drug in the skin (Foldvari et al. 1990).

The research on liposomal delivery systems, during the last two decades, has shown that the nature of lipids, size, charge, and even the method of preparation of liposomes could influence the dermatokinetics of drugs. One of the studies assessed the bioavailability of hydrocortisone, betamethasone, and triamcinolone acetonide from different liposomal formulations. In general, the liposomal formulations enhanced the drug delivery significantly over the ointment formulation. Interestingly, the liposomes prepared with stratum corneum lipids were found to be superior over phospholipid liposomes in terms of drug delivery efficiency (Fresta and Puglisi 1997).

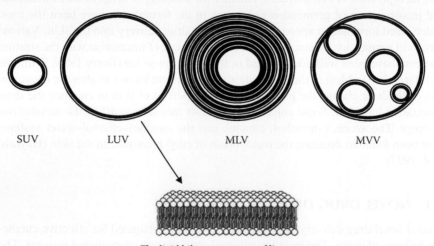

The lipid bilayer structure of liposomes

FIGURE 3.2 Schematic presentation of different types of liposomes. SUV, small unilamellar vesicles; LUV, large unilamellar vesicles; MLV, multilamellar vesicles; MVV, multivesicular vesicles.

Sentjurc and coworkers have shown that in case of smaller size liposomes (<200 nm), the delivery of hydrophilic drugs reduced significantly. This was because the smaller liposomes are not stable and disintegrate immediately in contact with the skin surface. They form a lipid layer over the skin, thus enhancing the barrier. However, in case of larger liposomes, some delivery into deeper skin layers was observed, which did not depend on the vesicle size significantly (Sentjurc et al. 1999). Contradicting this, some reports show "the smaller the better" concept to explain the inverse relationship between size and delivery efficiency of liposomal drug delivery systems (Lasch et al. 1992).

In most of the studies, positively charged liposomes enhanced the drug penetration relatively better than the negatively charged similar composition liposomes (Sinico et al. 2005). For example, the delivery of amphotericin B was found to be significantly higher in case of positively charged liposomes than the negatively and neutrally charged liposomes (Manosroi et al. 2004). The common mechanism put forth is that the positively charged liposomes are able to interact relatively well with the negatively charged skin surface at physiological pH conditions than anionic and nonionic liposomes.

3.3.1.2 Mechanism of Enhancement of Drug Delivery by Liposomes

The different mechanisms responsible for enhancement of drug delivery by liposomal formulations include free drug mechanism, skin permeability enhancement, vesicle adsorption or fusion with stratum corneum, intact vesicular penetration mechanism, and trans-appendageal penetration mechanism (El Maghraby et al. 2006, 2008) (Figure 3.3).

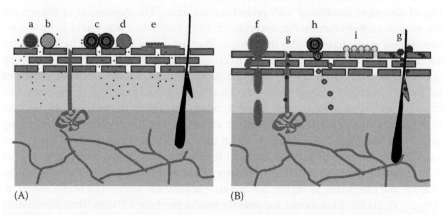

FIGURE 3.3 (See color insert.) (A) *Potential mechanisms of prolonged drug delivery using novel drug delivery systems*: The microparticles ((a) reservoir and (b) matrix type), liposomes ((c) MLV and (d) LUV) adhere onto the skin surface and slowly release the drug. (e) Lipid nanoparticles or liposomes form a barrier layer and decrease the rate of drug delivery into skin. (B) *Potential mechanisms of enhanced drug delivery or drug targeting using novel drug delivery systems*: (f) The ultradeformable liposomes penetrate the stratum corneum and enter deeper layers due to the gradient in the water content. (g) Enhanced permeation or targeting of drugs to skin appendages and furrows in the stratum corneum. (h) MLVs lose peripheral layers during penetration across the stratum corneum. (i) The vesicular drug delivery systems permeabilize the lipid domain by fusion or by release of enhancer, thus enhancing the drug penetration.

Several reports have clearly shown that the whole liposomes are not able to penetrate through the skin, but just lead to transfer of drug (Liu et al. 2004; Yu and Liao 1996). The most accepted mechanism is the free drug mechanism in which the release of drug from liposomes on the skin surface and permeation of free drug alone take place (Artmann et al. 1990; Ganesan et al. 1984). On the other hand, enhancement of skin permeability by the lipids was suggested as a potential mechanism for drug delivery enhancement in several studies (El Maghraby et al. 1999; Hofland et al. 1995; Kato et al. 1987). Kirjavainen and group have investigated this mechanism by resonance energy transfer studies and found that the lipids of liposomes penetrate into stratum corneum of skin by adhering to the surface followed by destabilization, fusion, or mixing with the lipid matrix of the stratum corneum, thus acting as penetration enhancers due to loosening of the lipid structure of stratum corneum (Kirjavainen et al. 1996).

Foldvari and coworkers have shown using electron micrographic pictures the presence of intact liposomes itself in the dermis region, suggesting the plausibility of penetration of intact liposomes into the skin layer. Interestingly, the liposomes seen in the dermis were unilamellar while multilamellar vesicles were applied to the skin. This was plausibly due to gradual rupture of the outer layers of liposomes during penetration through different layers of the skin (Foldvari et al. 1990; Fresta and Puglisi 1996; Natsuki et al. 1996).

So far, many studies have demonstrated the penetration of liposomes into the upper layers of the epidermis and into appendages of the skin. Liposomal formulations, which entered into appendageal routes, are recommended as potential vehicles for the treatment of disorders associated with pilosebaceous units. The deposition of fluorescent dyes such as carboxyfluorescein, and calcein, from liposomes was found to be significantly higher in the hair follicles than in the other regions of the skin, which was demonstrated using quantitative florescence microscopy analysis and confocal microscopy (Li et al. 1992; Lieb et al. 1992). Even the depth of penetration into the hair follicles was found to depend predominantly on the charge of the liposomes. Liposomes that are amphoteric and cationic were found to penetrate deeper in the hair follicles than the negatively charged liposomes (Jung et al. 2006). On the contrary, an interesting study by El Maghraby and group, in which they investigated the mechanism of liposomal delivery of estradiol, concluded that follicular pathway had a negligible role in the transdermal liposomal delivery of drug. They employed a sandwich of stratum corneum and epidermis as barrier for in vitro permeation studies (El Maghraby et al. 2001, 2006, 2008). This model supposedly blocks the hair follicles, thus eliminating the possibility of a shunt pathway.

3.3.1.3 Ultradeformable Vesicles

Ultradeformable vesicles consist of phospholipids, cholesterol, and in addition, a surfactant that acts as an edge activator. Transferosomes®, which are a form of ultradeformable liposomes, have been demonstrated to have the ability of squeezing through pores of less than one-tenth diameter of vesicles (Cevc et al. 1996; El Maghraby et al. 2006).

These unique vesicles are supposed to enhance the drug delivery across the skin only under non-occluded conditions. It has been proposed that the intact transferosomes are driven into skin by the force due to the transdermal hydration gradient. The hydrophilicity of phospholipids tends to avoid dry skin surface and thus in order to maintain swollen state, the vesicles follow the hydration gradient and penetrate into deeper skin layers (Cevc 1992; Cevc and Blume 1992; Cevc et al. 1995).

Cevc and coworkers assessed the transdermal penetrability of deformable liposomes. Compared to conventional gel, deformable liposomes resulted in 10-fold higher levels of diclofenac in the subcutaneous tissue with therapeutic levels delivered into the knee and even the systemic circulation (Cevc and Blume 2001, 2003, 2004). A study by Jain group, also reported a 12-fold enhancement in the bioavailability of zidovudine from elastic liposomes over the control formulation (Jain et al. 2006). Guo et al. have used elastic liposomes for the delivery of insulin while Gupta et al. used for cutaneous vaccination (Guo et al. 2000; Gupta et al. 2005a,b). In both the studies, the authors found better pharmacodynamic response in case of elastic liposomes over conventional liposomes.

3.3.1.4 Ethosomes

Ethosomes are multilamellar spherical vesicular systems made up of phospholipids, ethanol, proplyene glycol, and water. Generally, ethosomes have a vesicle size of about 50–200 nm. The enhancement of drug permeation through skin by ethosomes depends upon the size of the vesicles and their composition. The absorption of several drugs including minoxidil and testosterone was found to be many fold higher in case of ethosomal formulation as compared to liposomes (Touitou et al. 2000). The skin penetration enhancement of drugs by ethosomes was found to be due to their ability to disturb the lipid bilayer of the stratum corneum and resulted in an increased lipid fluidity (Lauer et al. 1996).

3.3.1.5 Niosomes

Niosomes are unilamellar or multilamelar vesicles similar to liposomes except that their ordered bilayer is made up of nonionic surfactants with or without cholesterol and with or without a charge-inducing agent. The fact that they are made up of nonionic surfactants makes them most suitable for cutaneous application. The two mechanisms likely to cause enhanced drug delivery are penetration of intact niosomes into upper layers of skin and modulation of lipid barrier in the stratum corneum.

3.3.2 Particulate Systems as Drug Carriers

3.3.2.1 Microparticulate Systems

Microparticulate systems are structures in which the drug is entrapped within the polymeric shell or matrix. The microparticles cannot penetrate the skin due to their size being larger than the penetration pathways in the skin. However, they are used in cutaneous delivery for longer retention and slower release of drugs in the applied region (Jelvehgari et al. 2006). Generally, the rate-controlling steps are the diffusion of drug from the particles into the external vehicle, dissolution of the polymeric shell to release the drug, or both. Microsponges are porous microspheres that are non-collapsible with the active ingredient entrapped in it (Chen et al. 2006; Embil and Nacht 1996; Jelvehgari et al. 2006; Wester et al. 1991).

3.3.2.2 Nanoparticulate Systems

Nanosystems contain particles in the nano sizes ranging from 10 to 1000 nm. Nanoparticles have been utilized as topical vehicles to prolong the residence time of drugs and biotherapeutics. Nanosystems possess higher specific surface area and provide homogeneous release of active substances into the skin. The polymers used in the preparation of polymeric nanoparticles include poly(ε-caprolactone), poly (lactide-co-glycolide), (poly(ε-caprolactone)-$block$-poly(ethylene glycol), poly(butyl cyanoacrylate), poly(ethyl cyanoacrylate), ethyl cellulose, cellulose acetate phthalate, chitosan, lipids, and fatty acid-conjugated poly(vinyl alcohol).

There is still ongoing debate regarding the penetrability of nanoparticles into the skin and the potential pathways involved there in. Alvarez-Roman and coworkers investigated the distribution and penetration pathways of fluorescent polymeric nanoparticles using confocal laser scanning microscopy in the porcine skin (Alvarez-Roman et al. 2004). The polymeric nanoparticles were found accumulated in the follicular openings and some nanoparticles were found to localize in the furrows of the skin.

Particularly, the lipid-based nanoparticles formulated as solid lipid nanoparticles (SLN) and nanostructured lipid carriers (NLC) have been proposed for a number of applications in topical drug delivery. SLN contain a solid particle matrix composition, which will distinguish them from NLC. SLN are prepared by replacing the liquid lipid of an o/w emulsion with solid lipids dispersed uniformly in the aqueous layer. The dispersion needs to be stabilized with a surfactant (0.5%–5% w/w if necessary). NLC are classified as second-generation lipid nanoparticles, which contains a mixture of solid lipid and liquid lipids. The lipid nanoparticles have been reported to enhance the permeation across the skin of several therapeutic agents, including macromolecules (Bhaskar et al. 2009; Larese et al. 2009; Mei et al. 2003; Zhao et al. 2010). The lipid nanoparticles have been shown to enhance the drug permeation owing to their ability to form an occlusive film on the skin, which will enable the skin to be in a hydrated condition.

Sanna et al, described the preparation and characterization of SLN-encapsulated econazole nitrate hydrogels. They used the tape-stripping method for the determination of drug retained in the skin. The release of drug from SLN depended on the lipid content in the nanoparticles. In vivo studies demonstrated that SLN promoted a rapid penetration of econazole through the stratum corneum after 1 h and improved the diffusion of drug in the deeper skin layers after 3 h of application compared with that of the control gel (Sanna et al. 2007).

3.3.2.3 Emulsified Systems: Micro- and Nanoemulsions

Microemulsions are optically clear systems (size of the dispersed phase globules ~300 nm) resembling (mixed) micelles, except that they contain an extra oily component. They form spontaneously and hence are highly stable systems. Nanoemulsions or submicron emulsions typically contain 20–500 nm size droplets stabilized by surfactants. Unlike microemulsions, they rely on energy input to form an emulsion. This makes nanoemulsions a metastable system.

3.4 SYSTEMIC DELIVERY APPROACHES

The drug is often administered via noncutaneous routes to treat cutaneous disorders. The systemically administered drugs reach the cutaneous tissues by natural physiological distribution. The distribution and elimination of drug in the skin tissue is determined by factors such as protein binding, binding of drug to the skin tissue, metabolic degradation due to specific enzymes present in the skin, molecular size, and physicochemical properties of the drug. A representative concentration profile graph of drug in skin extracellular fluid and blood plasma following intravenous administration is shown in Figure 3.4.

One of the most important factors that determines the drug distribution and elimination in cutaneous tissue following systemic administration is the health status of the patient. For example, Lanao and coworkers studied the concentration of amikacin in blister fluid in healthy volunteers and in patients with renal impairment following i.v. administration. The concentration of the antibiotic in serum and blister fluid was modified significantly in patients with renal impairment as compared to healthy volunteers. The penetration of amikacin into skin blister fluid in these patients increased progressively and a linear relationship was observed between the antibiotic in serum and blister fluid (Lanao et al. 1983).

The interstitial fluid compartment or the extracellular fluid is generally considered as a buffer fluid volume for the plasma. The extracellular fluid volume changes due to the administration of large volumes of liquids or osmotic agents into the plasma compartment. Skin is one such peripheral organ in which extracellular fluid reacts significantly for any fluid volume changes in the plasma. Larsson and Ware investigated the effect of isotonic fluid load on the plasma water and extracellular fluid volumes in the normal rats (Larsson and Ware 1983). They found that, following infusion

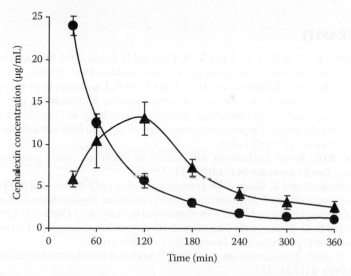

FIGURE 3.4 Concentration-time profile of cephalexin in skin following intravenous administration in rats. (•) Concentration of free drug in plasma. (▲) Concentration of free drug in the skin extracellular fluid.

of saline up to 130% of total plasma volume, the extracellular fluid volume in skin increased by 28%. Similarly, in another study, when high molecular weight dextran was infused, the extracellular fluid volume decreased due to fluid shift from interstitial compartment to plasma owing to high osmotic pressure. This kind of expansion and contraction in the extracellular fluid volume in the skin could influence the dermatokinetics of drugs significantly. In a study by our group, one group of rats was administered with dextran 75,000 intravenously and the other group was vehicle control. Twenty minutes after administration of dextran or vehicle, cephalexin solution was injected intravenously. The time course of drug in the skin extracellular fluid was investigated by sampling the drug from skin using cutaneous microdialysis technique. The C_{max} and $AUC_{0-6\,h}$ of cephalexin in the dextran group of rats was significantly twofold higher than the control (Unpublished data). Therefore, potential influence of any change in the fluid volume in the peripheral compartment on the drug pharmacokinetics should be considered seriously.

3.5 CONCLUSIONS

The dermatokinetics of drugs could be modified by exploiting simple physicochemical principles and small adjustments in the vehicle properties. Novel drug delivery systems such as vesicular systems and particulate systems could be developed to modulate the dermatokinetics of drugs. The type and composition of novel drug delivery systems should be opted and optimized to achieve specific objectives such as controlled release, targeted drug delivery, or enhanced drug penetration. Systemic drug delivery is one of the treatment options of cutaneous diseases due to the natural distribution of drug into cutaneous tissues. However, disease conditions and extracellular fluid volume changes could significantly influence the dermatokinetics of drugs.

BIBLIOGRAPHY

Alvarez-Roman, R., A. Naik, Y. N. Kalia, R. H. Guy, and H. Fessi. 2004. Skin penetration and distribution of polymeric nanoparticles. *J Control Release* 99 (1):53–62.

Artmann, C., J. Roding, M. Ghyczy, and H. G. Pratzel. 1990. Liposomes from soya phospholipids as percutaneous drug carriers. 1st communication: Qualitative in vivo investigations with antibody-loaded liposomes. *Arzneimittelforschung* 40 (12):1363–1365.

Barry, B. W. 1988. Action of skin penetration enhancers—The Lipid Protein Partitioning theory. *Int J Cosmet Sci* 10 (6):281–293.

Barry, B. W. 2001. Novel mechanisms and devices to enable successful transdermal drug delivery. *Eur J Pharm Sci* 14 (2):101–114.

Beall, H., R. Prankerd, and K. Sloan. 1994. Transdermal delivery of 5-fluorouracil (5-FU) through hairless mouse skin by 1-alkyloxycarbonyl-5-FU prodrugs: Physicochemical characterization of prodrugs and correlations with transdermal delivery. *Int J Pharm* 111 (3):223–233.

Beall, H. D. and K. B. Sloan. 1996. Transdermal delivery of 5-fluorouracil (5-FU) by 1-alkylcarbonyl-5-FU prodrugs. *Int J Pharm* 129 (1–2):203–210.

Benson, H. A. 2005. Transdermal drug delivery: Penetration enhancement techniques. *Curr Drug Deliv* 2 (1):23–33.

Bhaskar, K., J. Anbu, V. Ravichandiran, V. Venkateswarlu, and Y. M. Rao. 2009. Lipid nanoparticles for transdermal delivery of flurbiprofen: Formulation, in vitro, ex vivo and in vivo studies. *Lipids Health Dis* 8:6.

Calpena, A. C., C. Blanes, J. Moreno, R. Obach, and J. Domenech. 1994. A comparative in vitro study of transdermal absorption of antiemetics. *J Pharm Sci* 83 (1):29–33.

Cevc, G. 1992. Rationale for the production and dermal application of lipid vesicles. In *Liposome Dermatics*, eds H. C. Korting, H. I. Maibach, and O. Braun-Falco. Springer-Verlag, Berlin, Germany.

Cevc, G. and G. Blume. 1992. Lipid vesicles penetrate into intact skin owing to the transdermal osmotic gradients and hydration force. *Biochim Biophys Acta* 1104 (1): 226–232.

Cevc, G. and G. Blume. 2001. New, highly efficient formulation of diclofenac for the topical, transdermal administration in ultradeformable drug carriers, Transfersomes. *Biochim Biophys Acta* 1514 (2):191–205.

Cevc, G. and G. Blume. 2003. Biological activity and characteristics of triamcinolone-acetonide formulated with the self-regulating drug carriers, Transfersomes. *Biochim Biophys Acta* 1614 (2):156–164.

Cevc, G. and G. Blume. 2004. Hydrocortisone and dexamethasone in very deformable drug carriers have increased biological potency, prolonged effect, and reduced therapeutic dosage. *Biochim Biophys Acta* 1663 (1–2):61–73.

Cevc, G., G. Blume, A. Schätzlein, D. Gebauer, and A. Paul. 1996. The skin: A pathway for systemic treatment with patches and lipid-based agent carriers. *Adv Drug Deliv Rev* 18 (3):349–378.

Cevc, G., A. Schätzlein, and G. Blume. 1995. Transdermal drug carriers: Basic properties, optimization and transfer efficiency in the case of epicutaneously applied peptides. *J Control Release* 36 (1–2):3–16.

Chen, G., T. Sato, J. Tanaka, and T. Tateishi. 2006. Preparation of a biphasic scaffold for osteochondral tissue engineering. *Mater Sci Eng C Biomim Mater Sens Syst* 26 (1): 118–123.

Crommelin, D. and H. Schreier. 1994. Liposomes. In *Colloidal Drug Delivery Systems*, ed. J. Kreuter. CRC Press, New York.

El Maghraby, G. M., B. W. Barry, and A. C. Williams. 2008. Liposomes and skin: From drug delivery to model membranes. *Eur J Pharm Sci* 34 (4–5):203–222.

El Maghraby, G. M., A. C. Williams, and B. W. Barry. 1999. Skin delivery of oestradiol from deformable and traditional liposomes: Mechanistic studies. *J Pharm Pharmacol* 51 (10):1123–1134.

El Maghraby, G. M., A. C. Williams, and B. W. Barry. 2006. Can drug-bearing liposomes penetrate intact skin? *J Pharm Pharmacol* 58 (4):415–429.

El Maghraby, G. M., A. C. Williams, and B. W. Barry. 2001. Skin hydration and possible shunt route penetration in controlled estradiol delivery from ultradeformable and standard liposomes. *J Pharm Pharmacol* 53 (10):1311–1322.

Embil, K. and S. Nacht. 1996. The Microsponge Delivery System (MDS): A topical delivery system with reduced irritancy incorporating multiple triggering mechanisms for the release of actives. *J Microencapsul* 13 (5):575–588.

Foldvari, M., A. Gesztes, and M. Mezei. 1990. Dermal drug delivery by liposome encapsulation: Clinical and electron microscopic studies. *J Microencapsul* 7 (4):479–489.

Fresta, M. and G. Puglisi. 1997. Corticosteroid dermal delivery with skin-lipid liposomes. *J Control Release* 44 (2–3):141–151.

Fresta, M. and G. Puglisi. 1996. Application of liposomes as potential cutaneous drug delivery systems. In vitro and in vivo investigation with radioactively labelled vesicles. *J Drug Target* 4 (2):95–101.

Ganesan, M. G., N. D. Weiner, G. L. Flynn, and N. F. H. Ho. 1984. Influence of liposomal drug entrapment on percutaneous absorption. *Int J Pharm* 20 (1–2):139–154.

Guo, J., Q. Ping, and L. Zhang. 2000. Transdermal delivery of insulin in mice by using lecithin vesicles as a carrier. *Drug Deliv* 7 (2):113–116.

Gupta, P. N., V. Mishra, A. Rawat, P. Dubey, S. Mahor, S. Jain, D. P. Chatterji, and S. P. Vyas. 2005a. Non-invasive vaccine delivery in transfersomes, niosomes and liposomes: A comparative study. *Int J Pharm* 293 (1–2):73–82.

Gupta, P. N., V. Mishra, P. Singh, A. Rawat, P. Dubey, S. Mahor, and S. P. Vyas. 2005b. Tetanus toxoid-loaded transfersomes for topical immunization. *J Pharm Pharmacol* 57 (3):295–301.

Hayashi, T., Y. Iida, T. Hatanaka, T. Kawaguchi, K. Sugibayashi, and Y. Morimoto. 1997. The effects of several penetration-enhancers on the simultaneous transport and metabolism of ethyl nicotinate in hairless rat skin. *Int J Pharm* 154 (2):141–148.

Hofland, H. E., J. A. Bouwstra, H. E. Bodde, F. Spies, and H. E. Junginger. 1995. Interactions between liposomes and human stratum corneum in vitro: Freeze fracture electron microscopical visualization and small angle X-ray scattering studies. *Br J Dermatol* 132 (6):853–866.

Jain, S., A. K. Tiwary, and N. K. Jain. 2006. Sustained and targeted delivery of an anti-HIV agent using elastic liposomal formulation: Mechanism of action. *Curr Drug Deliv* 3 (2):157–166.

Jelvehgari, M., M. R. Siahi-Shadbad, S. Azarmi, G. P. Martin, and A. Nokhodchi. 2006. The microsponge delivery system of benzoyl peroxide: Preparation, characterization and release studies. *Int J Pharm* 308 (1–2):124–132.

Jona, J. A., L. W. Dittert, P. A. Crooks, S. M. Milosovich, and A. A. Hussain. 1995. Design of novel prodrugs for the enhancement of the transdermal penetration of indomethacin. *Int J Pharm* 123 (1):127–136.

Jung, S., N. Otberg, G. Thiede, H. Richter, W. Sterry, S. Panzner, and J. Lademann. 2006. Innovative liposomes as a transfollicular drug delivery system: Penetration into porcine hair follicles. *J Invest Dermatol* 126 (8):1728–1732.

Kato, A., Y. Ishibashi, and Y. Miyake. 1987. Effect of egg yolk lecithin on transdermal delivery of bunazosin hydrochloride. *J Pharm Pharmacol* 39 (5):399–400.

Kirjavainen, M., A. Urtti, I. Jaaskelainen, T. M. Suhonen, P. Paronen, R. Valjakka-Koskela, J. Kiesvaara, and J. Monkkonen. 1996. Interaction of liposomes with human skin in vitro— The influence of lipid composition and structure. *Biochim Biophys Acta* 1304 (3):179–189.

Lanao, J. M., A. S. Navarro, A. Dominguez-Gil, J. M. Tabernero, I. C. Rodriguez, and A. Gonzalez Lopez. 1983. Amikacin concentrations in serum and blister fluid in healthy volunteers and in patients with renal impairment. *J Antimicrob Chemother* 12 (5):481–488.

Larese, F. F., F. D'Agostin, M. Crosera, G. Adami, N. Renzi, M. Bovenzi, and G. Maina. 2009. Human skin penetration of silver nanoparticles through intact and damaged skin. *Toxicology* 255 (1–2):33–37.

Larsson, M. and J. Ware. 1983. Effects of isotonic fluid load on plasma water and extracellular fluid volumes in the rat. *Eur Surg Res* 15 (5):262–267.

Lasch, J., R. Laub, and W. Wohlrab. 1992. How deep do intact liposomes penetrate into human skin? *J Control Release* 18 (1):55–58.

Lauer, A. C., C. Ramachandran, L. M. Lieb, S. Niemiec, and N. D. Weiner. 1996. Targeted delivery to the pilosebaceous unit via liposomes. *Adv Drug Deliv Rev* 18 (3):311–324.

Li, L., L. B. Margolis, V. K. Lishko, and R. M. Hoffman. 1992. Product-delivering liposomes specifically target hair follicles in histocultured intact skin. *In Vitro Cell Dev Biol* 28A (11–12):679–681.

Lieb, L. M., C. Ramachandran, K. Egbaria, N. Weiner. 1992. Topical delivery enhancement with multilamellar liposomes via the pilosebaceous route: In vitro evaluations using fluorescent techniques with the hamster ear model. *J Invest Dermatol* 99:108–113.

Liu, H., W. S. Pan, R. Tang, and S. D. Luo. 2004. Topical delivery of different acyclovir palmitate liposome formulations through rat skin in vitro. *Pharmazie* 59 (3):203–206.

Manosroi, A., L. Kongkaneramit, and J. Manosroi. 2004. Stability and transdermal absorption of topical amphotericin B liposome formulations. *Int J Pharm* 270 (1–2):279–286.

Megrab, N. A., A. C. Williams, and B. W. Barry. 1995. Oestradiol permeation through human skin and silastic membrane: Effects of propylene glycol and supersaturation* 1. *J Control Release* 36 (3):277–294.

Megwa, S. A., S. E. Cross, H. A. E. Benson, and M. S. Roberts. 2000a. Ion-pair formation as a strategy to enhance topical delivery of salicylic acid. *J Pharm Pharmacol* 52 (8):919–928.

Megwa, S. A., S. E. Cross, M. W. Whitehouse, H. A. Benson, and M. S. Roberts. 2000b. Effect of ion pairing with alkylamines on the in-vitro dermal penetration and local tissue disposition of salicylates. *J Pharm Pharmacol* 52 (8):929–940.

Mei, Z., H. Chen, T. Weng, Y. Yang, and X. Yang. 2003. Solid lipid nanoparticle and microemulsion for topical delivery of triptolide. *Eur J Pharm Biopharm* 56 (2):189–196.

Mezei, M. and V. Gulasekharam. 1980. Liposomes—A selective drug delivery system for the topical route of administration. Lotion dosage form. *Life Sci* 26 (18):1473–1477.

Mezei, M. and V. Gulasekharam. 1982. Liposomes—A selective drug delivery system for the topical route of administration: Gel dosage form. *J Pharm Pharmacol* 34 (7):473–474.

Natsuki, R., Y. Morita, S. Osawa, and Y. Takeda. 1996. Effects of liposome size on penetration of dl-tocopherol acetate into skin. *Biol Pharm Bull* 19 (5):758–761.

Potts, R. O. and R. H. Guy. 1992. Predicting skin permeability. *Pharm Res* 9 (5):663–669.

Samad, A., Y. Sultana, and M. Aqil. 2007. Liposomal drug delivery systems: An update review. *Curr Drug Deliv* 4 (4):297–305.

Sanna, V., E. Gavini, M. Cossu, G. Rassu, and P. Giunchedi. 2007. Solid lipid nanoparticles (SLN) as carriers for the topical delivery of econazole nitrate: In-vitro characterization, ex-vivo and in-vivo studies. *J Pharm Pharmacol* 59 (8):1057–1064.

Sentjurc, M., K. Vrhovnik, and J. Kristl. 1999. Liposomes as a topical delivery system: The role of size on transport studied by the EPR imaging method. *J Control Release* 59 (1):87–97.

Sinico, C., M. Manconi, M. Peppi, F. Lai, D. Valenti, and A. M. Fadda. 2005. Liposomes as carriers for dermal delivery of tretinoin: In vitro evaluation of drug permeation and vesicle-skin interaction. *J Control Release* 103 (1):123–136.

Sloan, K. B. 1992. *Prodrugs: Topical and Ocular Drug Delivery*. Informa Healthcare, New York.

Stott, P. W., A. C. Williams, and B. W. Barry. 1998. Transdermal delivery from eutectic systems: Enhanced permeation of a model drug, ibuprofen. *J Control Release* 50 (1–3):297–308.

Stott, P. W., A. C. Williams, and B. W. Barry. 2001. Mechanistic study into the enhanced transdermal permeation of a model [beta]-blocker, propranolol, by fatty acids: A melting point depression effect. *Int J Pharm* 219 (1–2):161–176.

Touitou, E., N. Dayan, L. Bergelson, B. Godin, and M. Eliaz. 2000. Ethosomes—Novel vesicular carriers for enhanced delivery: Characterization and skin penetration properties. *J Control Release* 65 (3):403–418.

Valenta, C., U. Siman, M. Kratzel, and J. Hadgraft. 2000. The dermal delivery of lignocaine: Influence of ion pairing. *Int J Pharm* 197 (1–2):77–85.

Wester, R. C., R. Patel, S. Nacht, J. Leyden, J. Melendres, and H. Maibach. 1991. Controlled release of benzoyl peroxide from a porous microsphere polymeric system can reduce topical irritancy. *J Am Acad Dermatol* 24 (5 Pt 1):720–726.

Williams, A. C. 2003. *Transdermal and Topical Drug Delivery*. Pharmaceutical Press, London, U.K.

Williams, A. C. and B. W. Barry. 2004. Penetration enhancers. *Adv Drug Deliv Rev* 56 (5):603–618.

Yu, H. Y. and H. M. Liao. 1996. Triamcinolone permeation from different liposome formulations through rat skin in vitro. *Int J Pharm* 127 (1):1–7.

Zhao, X., Y. Zu, S. Zu, D. Wang, Y. Zhang, and B. Zu. 2010. Insulin nanoparticles for transdermal delivery: Preparation and physicochemical characterization and in vitro evaluation. *Drug Dev Ind Pharm* 36 (10):1177–1185.

4 Conventional Methods of Cutaneous Drug Sampling

José Juan Escobar-Chávez, Miriam López-Cervantes, and Adriana Ganem Rondero

CONTENTS

4.1 INTRODUCTION

The topical administration route is preferred to the systemic delivery to treat skin disorders as it provides dramatically higher skin-to-plasma ratios, maintaining therapeutically effective drug concentrations in the target organ with relatively less risk of inducing side effects due to high systemic exposure (Herkenne et al. 2008). The skin became popular as a potential site for systemic drug delivery, on the one hand, because of the possibility of avoiding the problems of stomach emptying, pH effects, enzyme deactivation associated with gastrointestinal passage, and hepatic first-pass metabolism, and, on the other hand, due to its capability to enable input control.

The most prevalent method for estimation of cutaneous drug delivery is still diffusion through excised animal/human skin or artificial membranes in the classical two-compartment Franz-type diffusion cells (Tegeder et al. 2002). Even though this in vitro system has proven to be a robust screening system for early qualitative prediction of bioavailability/bioequivalence (BE/BA), the method obviously has several limitations. Among the most critical are the lack of elimination routes in terms of the vascular system and viable metabolizing enzymes, alterations in the stratum corneum (SC) structure due to water uptake, and that the method actually determines percutaneous permeation instead of cutaneous penetration.

To obtain clinically relevant information about pharmacokinetic profiles in the skin, in vivo techniques must be applied. Previously, tape stripping (TS) of the skin has been a frequent technique to assess cutaneous drug delivery. This implies removal of the SC cell layers by consecutive adhesion of tape pieces to the skin surface and stripping of the top cell layers. The technique only assesses the penetration of drug into the SC (which is usually not the therapeutic target of cutaneous drug delivery) and can only determine a single concentration–time point per administration site.

Furthermore, indirect radiochemical methods, skin biopsies, pharmacodynamic methods, and, more rarely, suction blisters (which are usually applied to assess skin drug levels, following systemic administration) have also been applied to estimate in vivo skin penetration.

A number of excellent reviews that have been published contain detailed discussions concerning many aspects of the conventional methods of cutaneous drug sampling (Franz 1975, Touitou et al. 1998, Kreilgaard 2002, Lademann et al. 2008, Escobar-Chávez et al. 2008, 2009). The aim of this paper is to review the application of some of them: (i) the TS technique, (ii) skin biopsy (SB), and (iii) fluid blister sampling (FBS) techniques to investigate cutaneous drug delivery in animals and humans with focus on the variables associated with the method and the prediction of human cutaneous BE/BA from animal studies. This focus is justified by the magnitude of the experimental data available with the use of these techniques. The use of the TS, SB, and FBS techniques in experimental medicine and pharmaceutical sciences has a long history.

4.2 SKIN

The skin is the largest organ of the body (Potts et al. 1991, 1994, Forslind 1994), accounting for more than 10% of the body mass, and the one that enables the body to interact more intimately with its environment. The skin consists of four layers: the SC, is the outer layer of the skin (nonviable epidermis), and forms the rate-controlling barrier for diffusion for almost all compounds. It is composed of dead flattened, keratin-rich cells, the corneocytes. These dense cells are surrounded by a complex mixture of intercellular lipids, namely, ceramides, free fatty acids, cholesterol, and cholesterol sulfate. Their most important feature is that they are structured as ordered bilayer arrays (Ellias 1991). The predominant diffusional path for a molecule crossing the SC appears to be intercellular (Barry 1983, Hadgraft 2001, Walters et al. 2002, Guy et al. 2003). The other layers are the remaining layers of the epidermis (viable epidermis), the dermis, and the subcutaneous tissues. There are also several associated appendages: hair follicles, sweat ducts, apocrine glands, and nails. In a general context, the skin's functions may be classified as protective, homeostasis maintaining functions, or sensing (Nevill 1994).

Many agents are applied to the skin either deliberately or accidentally, with either beneficial or deleterious outcomes. The main interest in dermal absorption assessment is related to (i) local effects in dermatology (e.g., corticosteroids for dermatitis); (ii) transport through the skin seeking a systemic effect (e.g., nicotine patches and hormonal drug patches); (iii) surface effects (e.g., sunscreens, cosmetics, and antiinfectives) (Olvera-Martínez et al. 2005); (iv) targeting of deeper tissues (e.g., nonsteroidal anti-inflammatory agents) (Miyazaki et al. 1986, Chi et al. 1996, Kattan et al. 2000, Liaw et al. 2000, Curdy et al. 2001, Shin et al. 2001, Wang et al. 2001, Fang et al. 2002, Escobar-Chávez et al. 2005, 2006); and (e) unwanted absorption (e.g., solvents in the workplace, pesticides, or allergens) (Chao et al. 2004, Mattorano et al. 2004).

4.3 TAPE-STRIPPING TECHNIQUE

The simplest method for reducing the barrier imposed by the SC is to remove it. Theoretically, an adhesive tape removes a layer of corneocytes. This is a minimally invasive in vivo technique to sequentially remove SC by the repeated application of appropriate adhesive tapes (Pinkus 1951). In vivo, the removal of the SC by TS is performed by the repeated application of adhesive tapes to the skin's surface, which is shown in Figure 4.1. It has been found that on the flexor surface of the forearm, about 30 tape strips are needed to strip off most of the horny layer (Pinkus 1951). Multiple strips remove a substantial skin barrier, as evidenced by 20–25-fold increases in transepidermal water loss (TEWL) (Tsai et al. 1991). Usually, the amount of SC removed by TS is not linearly proportional to the number of tapes removed (Pinkus 1951). TS appears to be simple and easy to perform (Ohman et al. 1994, Sheth et al. 1998); however, there are different parameters that can influence the quantity of SC removed by a piece of tape, and these include TS mode (Tsai et al. 1991, Bashir et al. 2001), skin hydration, cohesion between cells (which increases with SC's depth), the body site, and inter-individual differences (King et al. 1979, Marttin et al. 1996). The impact of these factors has been frequently investigated (Pinkus 1951, King et al. 1979, Tsai et al. 1991, Ohman et al. 1994, Marttin et al. 1996, van der Molen et al. 1997, Sheth et al. 1998, Bashir et al. 2001, Löffler et al. 2004). A number of excellent reviews that have been published

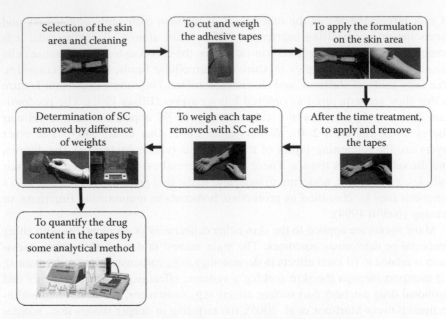

FIGURE 4.1 **(See color insert.)** Procedure of tape-stripping technique.

contain detailed discussions concerning many aspects of the TS technique (Pinkus 1966, Touitou et al. 1998, Moser et al. 2001, Lademann et al. 2008).

TS with adhesive tape is a widely used and accepted method to examine the localization and distribution of substances within the SC (Rougier et al. 1986, Tojo et al. 1989, Bommannan et al. 1990, Pershing et al. 1992, Higo et al. 1993, Lotte et al. 1993, Escobar-Chávez et al. 2005). This technique can be used to investigate SC cohesion in vivo by the amount of SC removed (King et al. 1979). Today, weighing with precision balances is the most frequently used method to determine the amount of SC removed on a tape strip. The method is also used to provide information about the kinetics of transdermal drug delivery, offering an apparently easy and quite non-invasive methodology for skin tissue sampling, and is the basis of the Food and Drug Administration (FDA's) so-called dermatopharmacokinetic (DPK) approach to the assessment of topical BA and BE (Shah 1998). However, validation and optimization of the procedure have not come quickly and the proposed guidance document has been withdrawn for reevaluation. More recent work has addressed at least some of the important limitations of the DPK approach (Conner 2001, Franz 2001, Pershing 2001) and has proposed modifications in order to incorporate it into an improved protocol.

After its description by Pinkus [8], TS has become a standard method in derma-tological research (Surber et al. 1999). This method can be used to obtain a more susceptible skin, for example, prior to the application of an irritant (Nangia 1993) or an allergen (van Voorst-Vader et al. 1991, Kondo et al. 1998, Surakka et al. 1999). Similarly, TS is performed to induce a defined disruption of the water barrier, for example, to evaluate the effect of a subsequently applied skin-care product in barrier restoration (Fluhr et al. 1999). It may be also used to obtain cells for mycological

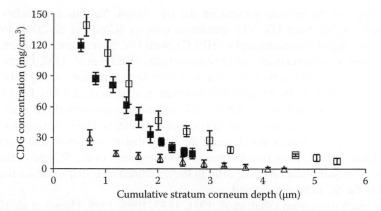

FIGURE 4.2 Concentration profiles for the Hibitane® solution in 70% ethanol by passive diffusion (Δ) (mean ± SD; n = 6) and iontophoresis (□) (mean ± SD; n = 4), and 0.5% clorhexidine digluconate aqueous solution (■) (mean ± SD; n = 2) applied by iontophoresis across human SC.

culture (Pechere et al. 1995, 1999) or to investigate SC quality (Ghadially et al. 1995). In dermatopharmacology, the SC barrier function (van der Valk et al. 1990, Fluhr et al. 1999) and the BA and BE of topical drugs (Sheth et al. 1987, Wilhelm et al. 1991, Hostýnek et al. 2001, Lademann et al. 2001) can be evaluated with the use of this technique (Shah et al. 1998, Schafer et al. 2001). Because of the limited systemic absorption of topical products, BA and/or BE studies used endpoint parameters or surrogate pharmacodynamic markers. However, if a drug does not penetrate or partition into SC, a pharmacodynamic activity will not take place. The skin-stripping methodology allows the determination of the uptake and elimination profile of topically applied drugs.

In the case of BE studies, topical BA can be estimated from the drug concentration within the SC, which is expected to be related to the drug concentration at the target site (i.e., usually viable epidermis or dermis) since SC is the rate-limiting barrier for percutaneous absorption. Similarly, to determine the drug concentration in blood and/or urine as surrogate for the concentration at the target tissue, the determination of the drug concentration in the SC may serve as a surrogate for the concentration in the viable (epi-)dermis (Schafer et al. 2001). A typical profile obtained from a skin permeation study with clorhexidine digluconate is shown in Figure 4.2. TS, which enables the removal of the SC layer by layer, is a useful DPK technique for the assessment of drug amounts in SC as a function of time (Shah et al. 1998).

4.3.1 Tape-Stripping Technique Protocol

Sequential TS of the SC allows horizontal fractions of the membrane to be obtained. The tape strips must then be extracted to recover and quantify the absorbed drug. Local BA may be assessed either from the combined tape strips or from the individual tape strips. The method of quantification depends, of course, on the nature of the drug and the amount present on the tape strips. Various approaches have been used ranging from UV–VIS spectrophotometry (Curdy et al. 2001) to high performance liquid chromatography (HPLC) (with UV, fluorescence, or even mass spectrophotometric detection), gas chromatography (Keiko et al. 1987, Goetz et al. 1988, Pragst et al. 2004), high performance thin layer chromatography (HPTLC) (Escobar-Chávez et al. 2005), scintillation counting (Caron et al. 1990, Bucks et al. 1993, Lücker 1984, Lücker et al. 1994), and infrared spectroscopy (Pirot et al. 1997, Reddy et al. 2002, Ayala-Bravo et al. 2003, Tsai et al. 2003). The key criteria are that the extraction process does not degrade the drug, that it is efficient and reproducible, and that it is free from interference from components of the SC and/or the tape adhesive. The quantification of the drug in the combined tape strips enables the total amount in the SC to be determined.

After much discussion (Shah et al. 1991, 1993, 1995, 1998, Huang et al. 1993, a draft guidance was published by the FDA in 1998 (Shah 1998) in which the general procedures for conducting a BA/BE study were described. The guidance proposed the following steps:

1. After application of a drug product for a given application time, excess formulation was to be removed by an appropriate procedure.
2. Two sequential tape strips of the SC were to be discarded.
3. Subsequently, 10 strips from the application site were to be made, combined, and extracted with a suitable solvent.
4. The entire process was to be performed at multiple sites following different times of application (Figure 4.3) (phase of absorption) and different periods between the longest application time and the time at which TS was performed (phase of elimination) (see Figure 4.4a and b).
5. The total amount of drug in the SC was then to be displayed as a function of time as a so-called DPK profile, characterized by a maximum amount, the time at which this maximum was achieved, and an area under curve (AUC).
6. Simultaneous comparison of two formulations of the same drug would allow an assessment of local BE using the representative DPK parameters.

Test treatment time (min)

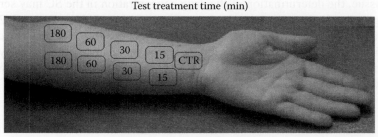

Reference treatment time (min)

FIGURE 4.3 (See color insert.) Assessment of topical BE of test and reference formulations. The SC is tape stripped immediately after each treatment time to determine the drug level in the barrier (CTR, control).

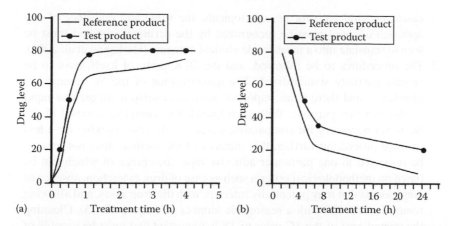

FIGURE 4.4 Representation of DPK study in the absorption phase (a) and elimination phase (b) to assess topical BE between test and reference formulations.

4.3.2 DISADVANTAGES AND UNSOLVED PROBLEMS OF TAPE-STRIPPING TECHNIQUE

Although useful as a jumping-off point, the guidance was open to a number of criticisms, and exposed some clear flaws. In brief, while many of these problems have now been recognized, but not fully resolved, the document has therefore been withdrawn, and, the possibility of a refined version being developed is under examination. Among the weaknesses of the original guidance, one may cite the following:

1. The timing of the TS procedures during the absorption and elimination phases is not clearly delineated.
2. The number of TS procedures is considerable, rendering a DPK evaluation onerous in terms of the work involved.
3. The unvalidated discarding of information from the first two tape strips has never been subjected to rigorous examination; indeed, it has been argued that these data are relevant to a complete BA assessment (Alberti et al. 2001).
4. Ten tape strips do not remove the same amount of SC from all subjects. As a result, drug will be measured in different volumes of the matrix, rendering comparisons between different treatments/subjects essentially meaningless.

Despite the clear value of a DPK approach based upon TS, there remains work to be done before its usefulness can be fully exploited:

1. The method has to be validated appropriately via clear demonstrations that significant differences in drug uptake into normal SC translates into clinically distinguishable scenarios in the real world; that is, the hypothesis that the SC concentration profile reflects drug availability at the site of action must be confirmed. Although it is clearly expedient to move quickly to an

easier approval path for generic topicals, the value and relevance of any approach adopted must be recognized by the dermatologist and must be seen to translate into a measurable clinical outcome (Herkenne et al. 2008).

2. The procedures to be followed, and the DPK protocol itself, have to be at least partially standardized. The quantification of the SC removed is mandatory and there is an important need to develop a simple and rapid method for this purpose. With such knowledge, complete removal of the SC is not necessary for comparative purposes, thereby allowing for a less invasive procedure; further, this means that the method does not have to be restricted to one particular adhesive tape, the choice of which can be based on methodological criteria such as ease of drug extraction, absence of components that may potentially interfere with the drug assay, and allowing removal of the SC with a reasonable number of strips (~10–15). Cleaning the treated area of the SC prior to TS is important and must be capable of removing excess formulation efficiently without inadvertently driving drug into the barrier. This includes displacement of vehicle from the natural furrows in the skin (Higo et al. 1993), the depth of which can be significantly greater than the SC thickness. Debate continues as to whether one or two tape strips should be taken initially and discarded, the argument being that any drug lodged on these layers, even after cleaning the skin, would never be absorbed. For the moment, no consensus exists on this point but clearly it will be necessary to better define the procedures to be followed when cleaning the skin at the end of the application period.

   This issue is particularly important both for poorly penetrating chemicals and for drug products administered in complex vehicles (such as those including liposomes or nanoparticles) (Herkenne et al. 2008).

3. The question of spatial localization of the drug must be addressed; specifically, it has been recognized that the target for certain drugs may be a skin appendage and that particular formulations have been proposed for the optimization of delivery to these structures. Exactly how TS might be used to compare vehicles that set out to achieve, for example, follicular targeting, has not been demonstrated. In this case, it would be necessary to show that SC levels are correlated with drug amounts in the appendage; however, if a formulation did specifically target a follicle, it might be logical to think that one would find lesser drug in the SC as a result. The problem is complex, therefore, and may need to call upon the application of recent work that has attempted to deduce the contribution of follicular transport to total drug delivery by comparing SC uptake in normal skin to that in skin whose follicles have been physically sealed (Lademann et al. 2003, 2004a, Toll et al. 2004, Teichmann et al. 2005). Obviously, further validation is required using drugs designed to act on these appendage structures (Herkenne et al. 2008).

Finally, there is the question of the relevance of the DPK approach, which has been developed for use on normal skin, to drug performance on diseased skin. In part, this relates to the validation issue discussed above and the need to correlate DPK measurements with clinical outcome. It is important to acknowledge, furthermore, that the methodology in its present state does not differentiate free and bound drug in the SC.

4.3.3 USE OF THE *TAPE-STRIPPING* TECHNIQUE TO DETERMINE THE PENETRATION OF SUBSTANCES THROUGHOUT THE SKIN

Removal by TS of the outermost skin layer, the SC, has become a common practice in recent years (Nylander-French et al. 2000, Surakka et al. 2000, Chao et al. 2004, Löffler et al. 2004, Mattorano et al. 2004, Escobar-Chávez et al. 2005). The determination of the kinetics and penetration depth of different kinds of permeants by tracing the concentration profiles in SC has been facilitated by the use of the virtually noninvasive method of SC stripping with adhesive tape (Rougier et al. 1986, Bommannan et al. 1990, Higo et al. 1993, van der Molen et al. 1997, Löffler et al. 2004). For this reason, TS also offers the possibility of evaluating BE of topical dermatological dosage forms (Rougier et al. 1986).

DPK characterization of active drugs in human volunteers has been suggested to be able to replace comparative clinical trials as a means of documenting BE (Shah et al. 1998). Moreover, in vitro methods are encouraged by regulatory agencies regarding the provision of percutaneous absorption data for drugs, pesticides, and cosmetics (Howes et al. 1996). All these points are emphasized in Table 4.1, which summarizes the research with the TS technique to determine the kinetics and penetration depth of permeants (drugs and toxic chemicals) (Park et al. 1995, van der Molen et al. 1997, Arima et al. 1998, Coderch et al. 1999, Curdy et al. 2001, Kalia et al. 2001, Lboutounne et al. 2002, Liu et al. 2002, Reddy et al. 2002, Wagner et al. 2002, Ayala-Bravo et al. 2003, Hara et al. 2003, Morgan et al. 2003, Padula et al. 2003, Tsai et al. 2003, Abdulmajed et al. 2004, Alvarez-Román et al. 2004, Chao et al. 2004, Jakasa et al. 2004, 2007, Jarvis et al. 2004, Lademann et al. 2004b, Lodén et al. 2004, Sarveiya et al. 2004, Strid et al. 2004, Bashir et al. 2005, Escobar-Chávez et al. 2005, Esposito et al. 2005, Fresno-Contreras et al. 2005, Inoue et al. 2005, 2006, Olvera-Martínez et al. 2005, de Jongh et al. 2006, Ganem-Quintanar et al. 2006, Herkenne et al. 2006, Hostýnek et al. 2006, Jacobi et al. 2006, Lundgren et al. 2006, Pellanda et al. 2006, Teichman et al. 2006, van der Merwe et al. 2006, Benfeldt et al. 2007, Brian et al. 2007, Herkenne et al. 2007, Mavon et al. 2007, Wagner et al. 2007, Dias et al. 2008, Sylvestre et al. 2008, Wiedersberg et al. 2008).

4.3.3.1 Kinetics and Penetration Depth of Drugs

4.3.3.1.1 Analgesic and Anti-Inflammatory Drugs

Arima et al. (1998) investigated the effect of hydroxypropyl-β-cyclodextrin (HP-β-CD) on the cutaneous penetration and activation of ethyl 4-biphenylyl acetate (EBA), a prodrug of the nonsteroidal anti-inflammatory drug 4-biphenylylacetic acid (BPAA), from hydrophilic ointment, using hairless mouse skin in vitro. When the hydrophilic ointment containing a complex of EBA with HP-β-CD was applied to full-thickness skin, HP-β-CD facilitated the penetration of EBA into the skin, the BPAA flux through the tape-stripped skin was greater than that through full-thickness skin,

TABLE 4.1
Research on the Tape-Stripping Technique as a Method to Determine Skin Penetration of Different Kind of Permeants

Research	Outcome	Author (Year)
(1) Kinetics and Penetration Depth of Drugs		
Analgesic and Anti-Inflammatory Drugs		
Effect of Azone® and Transcutol® on skin permeation of sodium naproxen formulated in PF-127 gels.	The combination of Azone and Transcutol in PF-127 gels enhanced sodium naproxen penetration, with up to twofold enhancement ratios compared with the formulation containing Transcutol only.	Escobar-Chávez et al. (2005)
Administration of piroxicam from a commercially available gel to human volunteers, both passively and under the application of an iontophoretic current.	The total amount of drug recovered in the SC post-iontophoresis by TS was significantly higher than that found following passive diffusion for each application time.	Curdy et al. (2001)
Effect of hydroxypropyl-P-cyclodextrin (HP-P-CD) on the cutaneous penetration and activation of EBA, a prodrug of nonsteroidal anti-inflammatory drug BPAA, from hydrophilic ointment, using hairless mouse skin in vitro.	The enhancing effect of HP-P-CD on the cutaneous penetration of EBA would be largely attributable to an increase in the effective concentration of EBA in the ointment.	Arima et al. (1998)
Production and characterization of MO dispersions as drug delivery systems for indomethacin.	Reflectance spectroscopy demonstrated that indomethacin incorporated into MO dispersions can be released in a prolonged fashion. TS experiments corroborated this finding.	Esposito et al. (2005)
Unilaminar films of Eudragit E-100 prepared from naproxen-loaded nanoparticles vs. conventional films.	In vivo penetration studies showed no statistical differences for the penetrated amount of naproxen across the SC and the depth of penetration for the two films. The films formulated from nanoparticle dispersions were shown to be effective for the transdermal administration of naproxen.	Ganem-Quintanar et al. (2006)
Investigation of pig ear skin as a surrogate for human skin in the assessment of topical drug bioavailability by sequential TS of the SC.	Pig ear skin ex vivo is promising as a tool for topical formulation evaluation and optimization.	Herkenne et al. (2006)
Determination of DPK parameters describing the rate and extent of delivery into the skin of Ibuprofen in the ventral forearms of human volunteers.	Prediction and experimental tests agreed satisfactorily suggesting that objective and quantitative information, to characterize topical drug bioavailability, can be obtained from this approach.	Herkenne et al. (2007)

TABLE 4.1 (continued)
Research on the Tape Stripping Technique as a Method to Determine Skin Penetration of Different Kind of Permeants

Research	Outcome	Author (Year)
Examination of the diffusion of copper through human SC in vivo following application of the metal as powder on the volar forearm for periods of up to 72 h.	Copper will oxidize and may penetrate the stratum corneum after forming an ion pair with skin exudates. The rate of reaction seems to depend on contact time and oxygen availability. A marked inter-individual difference was observed in baseline values and the amounts of copper absorbed.	Hostýnek et al. (2006)
Comparison of the bioavailability of ketoprofen in a photostabilized gel formulation without photoprotection using a new DPK TS model and an established ex vivo penetration method using human skin.	The comparison of the amount of ketoprofen in the skin after 45 min with the amount penetrated through the excised skin during 36 h, suggests a change in the thermodynamic activity of ketoprofen during exposure.	Lodén et al. (2004)
Penetration kinetics of SLs in *Arnica montana* preparations, by using a stripping method with adhesive tape and pig skin as a model.	Gel preparation showed a decrease in penetration rate, whereas the penetration rate of ointments remained constant over time. The total amount of SLs penetrated depends only on the kind of formulation and the SLs-content, but not on SLs composition or on the extraction agent used.	Wagner et al. (2006)
Penetration experiments investigating several incubation times with three different skin flaps, using the Saarbruecken penetration model and the lipophilic model drug flufenamic acid.	A direct linear correlation was found between the SC/water partition coefficients and the drug amounts penetrated into the SC for all time intervals tested.	Wagner et al. (2002)

Corticosteroids

Research	Outcome	Author (Year)
Effect of dose and application frequency on the penetration of TACA into human SC in vivo.	Considerable TACA amounts were retained within the SC, independently of dose and application frequency.	Pellanda et al. (2006)
Comparison of the in vivo BA of different topical formulations of BMV using the vasoconstrictor assay and the DPK method.	The DPK approach offers a reliable metric with which to quantify transfer of drug from the vehicle to the SC, and may be useful for topical BA and BE determinations.	Wiederisberg et al. (2008)

(continued)

TABLE 4.1 (continued)
Research on the Tape Stripping Technique as a Method to Determine Skin Penetration of Different Kind of Permeants

Research	Outcome	Author (Year)
Competition of chloride released from a Ag/AgCl cathode on the iontophoretic delivery of DM-P.	TS experiments confirmed the enhanced delivery of DM-P by iontophoresis relative to passive diffusion, with DM-P concentration greater inside the barrier postiontophoresis than that in the donor.	Sylvestre et al. (2008)

Disinfectants

Sustained bactericidal activity of chlorhexidine base–loaded poly(Є-caprolactone) NCs against *Staphylococcus epidermidis* inoculated onto porcine ear skin.	Topical application of chlorhexidine base–loaded positively charged NCs in an aqueous gel achieved a sustained release of bactericide against *Staphylococcus epidermidis* for at least 8 h.	Lboutounne et al. (2002)

Drugs for Keratinization Disorders

Design of an all-*trans* RA topical release system that modifies drug diffusion parameters in the vehicle and the skin, in order to reduce the systemic absorption and side effects associated with the topical application of the drug to the skin.	RA encapsulation not only prolongs drug release, but also promotes drug retention in viable skin.	Fresno-Contreras et al. (2005)

Anesthetics

Behavior of a skin bioadhesive film containing lidocaine in vitro and in vivo.	In vivo experiments with TS indicated that the presence of water during film application is essential to achieve not only the proper adhesion, but also an effective accumulation.	Padula et al. (2003)
Evaluation of the relationship between DMD sampling and the DPK method when employed simultaneously for BE investigations of topical formulations.	DMD sampling proved effective and variability analyses demonstrated the feasibility of BE studies in as little as 18 subjects.	Benfeldt et al. (2007)

Keratolytics

Keratolytic efficacy of topical preparations containing SA in humans by TS, and quantification of SC removal by protein analysis.	TS combined with protein analysis was sensitive in detecting the keratolytic effect of SA within hours of application.	Bashir et al. (2005)

TABLE 4.1 (continued)
Research on the Tape Stripping Technique as a Method to Determine Skin Penetration of Different Kind of Permeants

Research	Outcome	Author (Year)
Retinoids and Antioxidants		
Novel synthetic technique to synthesize the co-drug RA-AsA ester from all-*trans*-retinyl chloride (RA) and l- AsA suspended in ethanol at low temperature.	The data suggest the potential value of RA-AsA co-drug for treating damage to skin resulting from UV-induced production of free radicals.	Abdulmajed et al. (2004)
Aquaporine-3		
Glycerol replacement corrects each of the defects in AQP3-null mice.	The findings establish a scientific basis for the >200-year-old empirical practice of including glycerol in cosmetic and medicinal skin formulations due to its influence on water retention and the mechanical and biosynthetic functions of the SC.	Hara et al. (2003)
Antimicotics		
To determine the cutaneous bioavailability and hence to evaluate the bioequivalence of topically applied drugs in vivo.	Integration of the concentration profile over the entire SC thickness, that is, the AUC, provides a measure of the cutaneous BA and hence can be used to assess the BE of topically applied terbinafine	Kalia et al. (2001)
Antiviral Drugs		
Contribution of SC barrier and microvascular perfusion in determining dermal tissue levels of hydrophilic drugs (aciclovir and penciclovir) in vivo.	There was no relationship between fiber depth and the amount of drug dialyzed, which suggests free movement of antiviral drug on reaching the aqueous environment of the dermis.	Morgan et al. (2003)
Anti-Varicella Zoster Virus Nucleoside		
Determination of the in vitro dermal delivery of a new class of lipophilic, highly potent and uniquely selective anti-VZV nucleoside analog compared with aciclovir.	Topical delivery of these compounds is highly promising as a new first-line treatment for VZV infections.	Jarvis et al. (2004)

(continued)

TABLE 4.1 (continued)

Research on the Tape Stripping Technique as a Method to Determine Skin Penetration of Different Kind of Permeants

Research	Outcome	Author (Year)
	Vaccines	
Effect of CpG-ODN on the immune response to an antigen applied to tape-stripped mouse skin by evaluating the production of cytokines and Ig isotypes.	Administration of CpG-ODN through skin is a simple strategy for patients with diseases like atopic dermatitits, which is characterized by Th2-dominated inflammation.	Inoue et al. (2005, 2006)
Administration of HIV-1 DNA vaccine with cytokine-expressing plasmids to the skin of mice by a new topical application technique involving prior elimination of keratinocytes using TS.	Topical application is an efficient route for DNA vaccine administration and that the immune response may be induced by DNA plasmids taken in by DCs, Langerhans cells, or others such as APCs.	Liu et al. (2001)
	Other Kind of Permeants	
Effect of sucrose esters (sucrose oleate and sucrose laureate in water or Transcutol, TC) on the SC barrier properties in vivo. Impact of these molecules on the in vivo percutaneous penetration of 4-hydroxybenzonitrile (4-HB).	A combination of sucrose esters (oleate or laureate) and TC is able to temporarily alter the SC barrier properties, thereby promoting 4-HB penetration.	Ayala-Bravo et al. (2003)
Absorption of 4CP in humans using TS experiments to assess the conditions under which diffusion alters TS results.	Chemical concentrations in TSs can be affected by diffusion during TS, but with $t_{TS} < 0.2\ t_{lag}$ and an exposure time $> 0.3\ t_{lag}$, TS concentrations are not significantly affected by t_{TS}.	Reddy et al. (2002)
Development of a sensitive method for the determination of PEGs with different MWs in the human SC obtained by TS.	The method showed to be suitable for studying permeability in normal and impaired skin with respect to MW in the range of 150–600 Da.	Jakasa et al. (2004, 2007)
Dependence of permeability on molecular weight with different forms of barrier disruption.	Irrespectively of the form of barrier disruption, not only higher amounts, but also more varieties of chemicals (larger molecules) may penetrate into the skin in the presence of a compromised barrier, compared with normal skin.	Tsai et al. (2003)

TABLE 4.1 (continued)

Research on the Tape Stripping Technique as a Method to Determine Skin Penetration of Different Kind of Permeants

Research	Outcome	Author (Year)
Measurement of the status of skin surface hydration of uraemic patients with the corneometer and skin surface hydrometer, the functional capacity and the urea concentration in SC by TS technique, as well as the response of eccrine sweat gland to sudorific agent (0.05% pilocarpine HCL) in 18 age-matched hemodialysis patients and 10 healthy volunteers.	The functional abnormalities of eccrine sweat glands may be accounted for dry skin in uraemic patients at least in part, but there is no correlation between xerosis and pruritus.	Park et al. (1995)
Study of the higher alkanols (hexanol, octanol, decanol) in vivo using a combination of ATR-FTIR spectroscopy and TS	In general, solvent uptake was proportional to the induced shift in the C–H stretching frequency. Lipid disorder was induced by all vehicles studied in vivo and was proportional to the amount of vehicle present in the skin.	Dias et al. (2008)
UV Absorbers		
Penetration of OMC encapsulated in poly(Є-caprolactone) nanoparticles into and across porcine ear skin in vitro.	Nanoparticulate encapsulation of OMC increased its "availability" within the SC.	Alvarez-Román et al. (2004)
In vivo distribution profile of OMC contained in NCs through the SC. Comparison with a NE and a conventional o/w EM.	NE increased the extent of OMC penetration relative to the penetration achieved by NCs or EM.	Olvera-Martínez et al. (2005)
Quantification of four common sunscreen agents, namely, 2-hydroxy-4 methoxybenzophenone, 2-ethylhexyl-*p*-methoxycinnamate, 2-ethylhexylsalicylate (octylsalicylate), and SA 3,3,5-trimethcyclohexyl ester in a range of biological matrices.	A preliminary clinical study demonstrates a significant penetration of all sunscreen agents into the skin, as well as of oxybenzone and its metabolites across the skin.	Sarveiya et al. (2004)

(*continued*)

TABLE 4.1 (continued)
Research on the Tape Stripping Technique as a Method to Determine
Skin Penetration of Different Kind of Permeants

Research	Outcome	Author (Year)
Amount of sunscreen present on the skin of people at the beach.	The best protected areas were the upper arm and décolleté, but even in these areas, most volunteers had applied only 10% of the COLIPA standard amount.	Lademann et al. (2004)
Penetration of TiO$_2$ and MBBT, included in a broad-spectrum sunscreen formulation, into human skin in vivo, using the TS method, and in vitro, using a compartmental approach.	In vivo and in vitro penetration studies showed an absence of TiO$_2$ penetration into viable skin layers through either transcorneal or transfollicular pathways, and a negligible transcutaneous absorption of MBBT.	Mavon et al. (2007)

Fragrances

In vitro human skin permeation and distribution of GN.	Systemic exposure resulting from the use of GN as a fragrance ingredient, under unoccluded conditions, would be low based on the currently reported use levels.	Brian et al. (2007)

Dyes

Development of a method to investigate the effectiveness of reservoir closure by different formulations. Model penetrant: Patent Blue V.	Application of barrier creams cannot replace other protective measures and should be maximally used to inhibit low-grade irritants or in combination with other protectants, or in body areas where other protective measures are not applicable.	Teichmann et al. (2006)
Penetration of highly hydrophilic and lipophilic dyes into the skin using pure oil or water, comparing them with an o/w EM.	Differences in the distribution and the localization of both dyes within the SC were observed. These differences depend on the physicochemical properties of both the vehicles and the dyes.	Jacobi et al. (2006)
Study of the reservoir capacity of the SC by topical application of sodium fluorescein encapsulated in vesicles in order to elucidate the mechanism involved in the (trans)dermal transport of drugs when vesicles are applied to the skin.	Similar values obtained when the same liposomes with varying encapsulation content were applied could lend support to the penetration mechanism where the vesicle enhancement is mainly due to SC structural modification.	Coderch et al. (1999)

TABLE 4.1 (continued)
Research on the Tape Stripping Technique as a Method to Determine Skin Penetration of Different Kind of Permeants

Research	Outcome	Author (Year)
(2) Dermal Absorption of Toxic or Irritant Chemicals		
Development and testing of a simple, noninvasive dermal sampling technique on human volunteers under laboratory conditions to estimate acute dermal exposure to jet fuel (JP-8).	Naphthalene has a short retention time in the human SC and the TS method, if used within 20 min of the initial exposure, can be employed to measure the amount of naphthalene in the SC due to a single exposure to jet fuel.	Mattorano et al. (2004)
Normalization of extracted concentrations of naphthalene (as a marker for jet fuel exposure) from human volunteers, before and after exposure to jet fuel (JP-8). Removal and quantification of keratin by SC-TS.	The amount of keratin removed with tape strips was not affected by an exposure of up to 25 min to JP-8, and there was a substantial decrease in the amount of keratin removed with consecutive TSs from the same site; thus, adjusting the amount of naphthalene to the amount of keratin measured in a tape strip sample should improve the interpretation of the amount of this analyte by using this sampling approach.	Chao et al. (2004)
Description of a PBPK model developed to simulate the absorption of organophosphate pesticides, such as parathion, fenthion, and methyl parathion, through porcine skin with flow-through cells.	The study demonstrated the utility of PBPK models for studying dermal absorption, which can be useful as explanatory and predictive tools.	Van der Merwe et al. (2006)
Study of whether the SLS penetration rate into the SC is related to an impairment of skin's water barrier function and inflammation.	Variation in barrier impairment and inflammation of human skin depends on SLS penetration rate, which was mainly determined by SC thickness.	De Jongh et al. (2006)
Modification and testing of a vacuuming sampler for removing particles from the skin.	Agreement between the vacuuming sampler and the TS technique.	Lundgren et al. (2006)
Development of a noninvasive sampling method for measuring dermal exposure to a multifunctional acrylate employing TS.	No significant difference was observed in recovery between TPGDA and UV-resin for the first TS when calculated as a percentage of the theoretical amount. TS can be used to quantify dermal exposure to multifunctional acrylates.	Nylander-French (2000)

while the activation of the prodrug in the skin was slowed by TS. Their results suggest that the enhancing effect of HP-β-CD on the cutaneous penetration of EBA would be largely attributed to an increase in the effective concentration of EBA in the ointment.

Curdy et al. (2001) administered piroxicam from a commercially available gel to human volunteers, both passively and under the application of an iontophoretic current. After treatment, the SC at the site of application was progressively tape-stripped and piroxicam transport into the membrane was assessed by UV-analysis of drug extracted from the tape strips. Current application enhanced drug uptake into the SC, as indicated by both increased piroxicam concentrations in the horny layer and detectable concentrations at greater depths in the membrane. The total amount of drug recovered in the SC post-iontophoresis was significantly higher than that found following passive diffusion for each application time.

Escobar-Chávez et al. (2005) determined the penetration of sodium naproxen, formulated in Pluronic F-127 gels containing Azone and Transcutol as penetration enhancers, through human skin in vivo by using the TS technique. It was found that the combination of Azone and Transcutol in PF-127 gels enhanced sodium naproxen penetration, with up to twofold enhancement ratios compared with the formulation containing only Transcutol. These results were confirmed by TEWL and ATR-FTIR (attenuated total reflectance-Fourier transform infrared) spectroscopy, suggesting a synergistic action for Azone and Transcutol.

Esposito et al. (2005) produced and characterized monoleine (MO) dispersions as drug delivery systems for indomethacin. An in vitro diffusion study was conducted using Franz cells associated to SC epidermal membrane on cubosome dispersions viscosized by carbomer. In vivo studies based on skin reflectance spectrophotometry and TS were performed to better investigate the performance of cubosome as an indomethacin delivery system.

Indomethacin incorporated in viscosized MO dispersions exhibited a lower flux with respect to the analogous formulation containing the free drug in the aqueous phase and to the control formulation based on carbomer gel. Reflectance spectroscopy demonstrated that indomethacin incorporated into MO dispersions can be released in a prolonged fashion. TS experiments corroborated this finding. MO dispersions can be proposed as nanoparticulate systems able to control the percutaneous absorption of indomethacin.

Ganem-Quintanar et al. (2006) used naproxen-loaded nanoparticles to prepare, in a one-step process, unilaminar films of Eudragit E-100. Nanoparticle films and conventional films were characterized in vitro by drug release studies through a cellulose membrane using Franz-type cells, and in vivo by penetration experiments with the TS technique. Concerning in vivo penetration studies, no statistical differences were found for the amount of naproxen penetrated across the SC and the depth of penetration for the two films.

Herkenne et al. (2006) investigated pig ear skin as a surrogate for human skin in the assessment of topical drug bioavailability by sequential TS of the SC. Ex vivo experiments on isolated pig ears were compared with in vivo studies in human volunteers. Four formulations including ibuprofen in different propylene glycol (PG)–water mixtures (25:75, 50:50, 75:25, and 100:0), were compared. Derived DPK parameters characterizing the diffusion and partitioning of the drug in the SC ex vivo were consistent with those in vivo following a 30 min application period. Furthermore, non-steady-state ex vivo results could be used to predict the in vivo concentration profile of the drug across the SC when a formulation was administered for 3 h (i.e., close to steady state). Taken together, the results obtained suggest that pig ear skin ex vivo is promising as a tool for topical formulation evaluation and optimization.

Continuing with their research, Herkenne et al. (2007) explored the potential of using SC-TS, post application of a topical drug formulation, to derive DPK parameters describing the rate and extent of delivery into the skin of ibuprofen in the ventral forearms of human volunteers for periods ranging between 15 and 180 min. Subsequently, SC was tape stripped, quantified gravimetrically, and extracted for drug analysis. Together with concomitant TEWL measurements, SC concentration–depth profiles of the drug were reproducibly determined and fitted mathematically. The SC-vehicle partition coefficient (K) and a first-order rate constant related to ibuprofen diffusivity in the membrane (D/L^2, where L ¼SC thickness) were derived from data fitting and characterized the extent and rate of drug absorption across the skin. Integration of the concentration profiles yielded the total drug amount in the SC at the end of the application period. Using K and D/L^2 obtained from the 30 min exposure, it was possible to predict ibuprofen uptake as a function of time into the SC. Prediction and experiment agreed satisfactorily suggesting that objective and quantitative information, with which to characterize topical drug BA, can be obtained from this approach.

Hostýnek et al. (2006) shed light on the long-standing controversy on whether wearing copper bangles benefits patients suffering from inflammatory conditions such as arthritis. Sequential TS was performed on healthy volunteers to examine the diffusion of copper through human SC in vivo, following application of the metal as powder on the volar forearm for periods of up to 72 h. Exposure sites were stripped 20 times, and the strips were analyzed for metal content by inductively coupled plasma-mass spectroscopy.

The results indicate that, in contact with skin, copper will oxidize and may penetrate the SC after forming an ion pair with skin exudates. The rate of reaction seems to depend on contact time and oxygen availability. A marked inter-individual difference was observed in baseline values and amounts of copper absorbed.

Lodén et al. (2004) compared the bioavailability of ketoprofen in a photostabilized gel formulation without photoprotection using a new DPK TS model and an established ex vivo penetration method using human skin. Analyses of the SC showed that during the first 45 min, about 12 µg/cm^2 of ketoprofen were absorbed into the skin from the formulations. The area under the ketoprofen concentration–time curve (AUC$_{0-6h}$) for the photostabilized gel/transparent gel ratio was 73%. The rate of penetration of ketoprofen through isolated skin was approximately 0.2 µg/cm^2 h for both formulations. The ratio's AUC$_{0-36h}$ was 84%. Thus, the two methods did not disagree in terms of the relative efficacy of the two gels. The comparison of the amount of ketoprofen in the skin after 45 min with the amount penetrated through the excised skin during 36 h suggests a change in the thermodynamic activity of ketoprofen during exposure. A supersaturated formulation may have been formed initially due to evaporation of ethanol.

Wagner et al. (2007) studied the penetration kinetics of sesquiterpene lactones (SLs) in *Arnica montana* preparations; a stripping method with adhesive tape and pig skin as a model was used. For the determination of SLs in the stripped layers of the SC, a gas chromatography/mass spectrometry method was developed and validated.

The penetration behavior of one gel preparation and two ointment preparations was investigated. The SLs of all preparations showed a comparable penetration and permeation through the SC, in the uppermost layer of the skin. Interestingly, the gel preparation showed a decreased penetration rate over 4 h, whereas the penetration rate of ointments remained constant over time. Moreover, they could demonstrate that the total amount of SLs penetrated depends only on the kind of formulation and the SLs content in the formulation, but not on SLs composition or the extraction agent used.

Wagner et al. (2002) carried out penetration experiments to investigate several incubation times with three different skin flaps using the Saarbruecken penetration model and the lipophilic model drug flufenamic acid. Drug distribution within SC was obtained by the TS technique, while the drug present in deep skin layers was determined by cryosectioning. In addition, for the lipophilic drug flufenamic acid, a direct linear correlation was found between SC/water partition coefficients and the drug amounts penetrated into the SC for all the time intervals tested. The authors concluded that SC/water-partition coefficients offer the possibility to predict drug amounts within the SC of different donor skin flaps, without a time-consuming determination of the lipid composition of the SC.

4.3.3.1.2 Corticosteroids

The aim of Pellanda et al. (2006) was to investigate the effect of (i) dose and (ii) application frequency on the penetration of triamcinolone acetonide (TACA) into human SC in vivo. The experiments were conducted on the forearms of 15 healthy volunteers, with (i) single TACA doses (300 and 100 µg/cm^2) and (ii) single (1 × 300 µg/cm^2) and multiple (3 × 100 µg/cm^2) TACA doses. SC samples were collected by TS after (i) 0.5, 4, and 24 h and (ii) 4, 8, and 24 h. In Experiment 1, TACA amounts within SC after application of 1 × 300 µg/cm^2 compared to 1 × 100 µg/cm^2 were only significantly different immediately after application, and were similar at 4 and 24 h. In (ii), multiple applications of 3 × 100 µg/cm^2 yielded higher TACA amounts compared to a single application of 1 × 300 µg/cm^2 at 4 and 8 h. At 24 h, no difference was observed. In conclusion, by using this simple vehicle, considerable TACA amounts were retained within the SC, independently of dose and application frequency.

Wiedersberg et al. (2008) compared the in vivo BA of different topical formulations of betamethasone 17 valerate (BMV) using the vasoconstrictor assay and the DPK method. BMV at the same thermodynamic activity in different vehicles provoked similar skin blanching responses, while DPK profiles distinguished between the formulations. Further, skin blanching responses and drug uptake into the SC clearly depended upon the absolute BMV concentration applied. However, while the saturable nature of the pharmacodynamic response was clear, the TS method distinguished unequivocally between the different formulations and different concentrations. The DPK approach offers a reliable metric with which to quantify transfer of drug from the vehicle to the SC and may be useful for topical BA and BE determinations.

Sylvestre et al. (2008) studied the competition of chloride released from a Ag/AgCl cathode on the iontophoretic delivery of dexamethasone phosphate (DM-P). Iontophoresis of DM-P was performed by using pig skin. A 0.3 mA constant current was applied via Ag/AgCl electrodes. The amounts of DM-P and dexamethasone (DM) were also quantified in the SC, using TS, after passive and iontophoretic delivery. The iontophoretic delivery of DM-P from pure water was unaffected by the accumulation of Cl⁻ released by the donor cathode when the drug's concentration was 4.25–17 mM. At 0.85 mM, however, Cl⁻ competition was significant and the drug flux was significantly reduced. The formulation of the drug in the presence of Cl⁻ resulted in a nonlinear dependence of flux on the molar fraction of the drug. TS experiments confirmed the enhanced delivery of DM-P by iontophoresis relative to passive diffusion, with DM-P concentration being higher inside the barrier postiontophoresis than that in the donor.

4.3.3.1.3 Disinfectants

Lboutounne et al. (2002) investigated the sustained bactericidal activity of chlorhexidine base–loaded poly(Є-caprolactone) nanocapsules (NCs) against *Staphylococcus epidermidis* inoculated onto porcine ear skin. The antimicrobial activity of these colloidal carriers was evaluated (i) in vitro against eight strains of bacteria and (ii) ex vivo against *Staphylococcus epidermidis* inoculated for 12 h onto porcine ear skin surface treated for 3 min either with 0.6% chlorhexidine base–loaded or unloaded NCs suspended in hydrogel, or 1% chlorhexidine digluconate aqueous solution. Chlorhexidine absorption into the SC was evaluated by the TS technique. The results showed that chlorhexidine NCs in aqueous suspension with a 200–300 nm size and a positive charge exhibited similar minimum inhibitory concentrations against several bacteria, compared with chlorhexidine digluconate aqueous solution. Ex vivo, there was a significant reduction in the number of colony forming units from skin treated with chlorhexidine NCs suspension for 3 min compared to chlorhexidine digluconate solution after an 8 h artificial contamination. Interestingly, NCs were present in porcine hair follicles. The topical application of chlorhexidine base–loaded positively charged NCs in an aqueous gel achieved a sustained release of bactericide against *Staphylococcus epidermidis* for at least 8 h.

4.3.3.1.4 Drugs for Keratinization Disorders

Fresno-Contreras et al. (2005) designed an all-*trans* retinoic acid (RA) topical release system that modifies drug diffusion parameters in the vehicle and the skin in order to reduce systemic absorption and side effects associated with the topical application of the drug to the skin. RA, either in free form or encapsulated in SC lipid liposomes, was included in hydrogels prepared with Carbopol® Ultrez™ 10 and hyaluronic acid. In vitro permeability experiments with [³H]-*t*-RA were carried out using a Franz-type diffusion cell in abdominal rat skin samples. The accumulation of the drug in the surface and skin layers was evaluated by both the TS technique and a dissection technique. The results show that RA encapsulation not only prolongs drug release, but also promotes drug retention in viable skin. At the same time, interaction between RA and hyaluronic acid has an obstructive effect on diffusion, which contributes to the formation of a reservoir.

4.3.3.1.5   Anesthetics

Padula et al. (2003) studied the behavior of a skin bioadhesive film containing lidocaine, in vitro and in vivo. Film characterization included in vitro and in vivo drug transport studies with and without iontophoresis. The release rate was compared with a lidocaine commercial gel. The permeation kinetics across the skin was not linear, but the patch acted as a matrix controlling drug delivery. Additionally, permeation rate increased with drug loading. The in vivo experiments with TS indicated that the presence of water during film application is essential to achieve not only the proper adhesion, but also an effective accumulation. The application of an electric current to the patch can further increase the amount of drug accumulated in the SC.

Benfeldt et al. (2007) evaluated the relationship between dermal microdialysis (DMD) sampling and the DPK method when employed simultaneously for BE investigations of topical formulations. Topical lidocaine cream and ointment (both 5%) was investigated in eight healthy human volunteers (four males, four females). On one forearm, four microdialysis probes in two penetration areas sampled for 5 h, and on the other, TS was performed 30 and 120 min after product application. Lidocaine content in samples was analyzed by HPLC–mass spectrometry. The two methods were in agreement showing three- to fivefold higher lidocaine penetration from cream formulation than from ointment. A rank-order correlation between the two methods was demonstrated for lidocaine contents in microdialysates versus the tape strip at 120 min, significant for the ointment formulation and for both formulations analyzed together. The analysis of variance demonstrated reproducible lidocaine concentrations in microdialysates with an intrasubject variability of 19% between probes and 20% between the two penetration areas. Thus, intersubject variability accounted for 61% of the variance. DMD sampling proved effective and variability analyses demonstrated the feasibility of BE studies in as little as 18 subjects.

4.3.3.1.6   Keratolytics

Bashir et al. (2005) studied the keratolytic efficacy of topical preparations containing salicylic acid (SA) in humans by the TS technique, quantifying SC removal by protein analysis. In combination with TS, squamometry was used to evaluate the influence of SA on skin surface scaliness and desquamation. Furthermore, skin barrier perturbation and skin irritability were recorded and related to the dermatopharmacological effect of the preparations. In contrast to squamometry, TS combined with protein analysis was sensitive in detecting the keratolytic effect of SA within hours of application. Importantly, whereas the pH of the preparations had only a minimal influence on efficacy, local dermatotoxicity was significantly increased at an acidic pH.

This indicates that the intent to increase the amount of free, non-dissociated SA is, in fact, counterproductive, as more acidic preparations resulted in skin irritation and barrier disruption.

4.3.3.1.7   Retinoids and Antioxidants

Abdulmajed et al. (2004) used a novel synthetic technique to synthesize the co-drug retinyl ascorbate (RA-AsA) ester from all-*trans*-retinyl chloride (RA) and L-ascorbic acid (AsA) suspended in ethanol at low temperature. The flux and permeation coefficient were determined using heat-separated human skin membrane, and skin penetration was determined by TS using full-thickness human skin. All experiments were performed in parallel with retinyl palmitate and ascorbyl palmitate. Overall, the data suggest the potential value of RA-AsA co-drug for treating damage to skin resulting from the UV-induced production of free radicals.

4.3.3.1.8 Aquaporine-3

Hara et al. (2003) showed that glycerol replacement corrects each of the defects in aquaporine-3 (AQP3)-null mice. SC water content, measured by skin conductance and ^3H_2O accumulation, was threefold lower in AQP3-null versus. wild-type mice, but was similar after topical or systemic administration of glycerol in amounts that normalized glycerol content in the SC. Orally administered glycerol fully corrected reduced skin elasticity in AQP3-null mice, as measured by the kinetics of skin displacement after suction, and the delayed barrier recovery, as measured by TEWL after TS. The analysis of [^{14}C] glycerol kinetics indicated a reduced blood-to-SC transport of glycerol in AQP3-null mice, resulting in slowed lipid biosynthesis. These data provide functional evidence for a physiological role of glycerol transport by aquaglyceroporin, and indicate that glycerol is a major determinant of SC water retention and of mechanical and biosynthetic functions. Their findings establish a scientific basis for the >200-year-old empirical practice of including glycerol in cosmetic and medicinal skin formulations.

4.3.3.1.9 Antimicotic Drugs

Kalia et al. (2001) presented a method to determine the cutaneous bioavailability and hence to evaluate the bioequivalence of topically applied drugs in vivo. The procedure uses serial TS and TEWL measurements to quantify the thickness of the removed SC and to determine the intact membrane thickness. Following TS, the drug is extracted from the tapes and assayed by HPLC. This provides the drug concentration profile of terbinafine as a function of the normalized position within the SC. The data are fitted to a solution of Fick's second law of diffusion in order to calculate characteristic membrane transport parameters. The integration of the concentration profile over the entire SC thickness, that is, the AUC, provides a measure of the cutaneous bioavailability and hence can be used to assess the bioequivalence of topically applied drugs.

4.3.3.1.10 Antiviral Drugs

Morgan et al. (2003) measured the contribution of SC barrier and microvascular perfusion in determining dermal tissue levels of two hydrophilic drugs (aciclovir and penciclovir) in vivo. Removal of the SC by TS resulted in a 1300-fold increase in penciclovir absorption and a 440-fold increase in aciclovir absorption, confirming that SC is the major barrier to hydrophilic drug absorption.

4.3.3.1.11 Anti-Varicella Zoster Virus Nucleoside

Jarvis et al. (2004) determined the in vitro dermal delivery of a new class of lipophilic, highly potent, and uniquely selective anti-Varicella Zoster virus (VZV) nucleoside analogue compared with aciclovir. Three test compounds (Cf1698, Cf1743, and Cf1712) and aciclovir were formulated in PG/aqueous cream, and finite doses were applied to full-thickness pig ear skin for 48 h in vertical Franz-type diffusion cells. Depth profiles were constructed following TS and membrane separation. All three test compounds reached the target basal epidermis in concentrations suggesting they would be highly efficacious in reducing the viral load. Furthermore, the data showed that each of the test compounds would have a far superior performance than aciclovir. The dermatomal site of viral replication during secondary infection—the basal epidermis—was successfully targeted.

4.3.3.1.12 Vaccines

The skin-associated lymphoid tissue, formed by powerful antigen-presenting cells (APCs), such as Langerhans cells (LCs), dermal dendritic cells (DCs), recirculating T cells, and regional LCs, ensures the efficient presentation of antigen to immuno-competent cells and the induction of strong immune responses. LCs and dermal DCs commonly exist in the skin and are easy to target (Strid et al. 2004). The TS technique has been used to study the effect of oligodeoxynucleotides on the immune response (Inoue et al. 2005) and expression of immune receptors (Inoue et al. 2007).

Inoue et al. (2005) examined the effect of CpG-oligodeoxynucleotide (CpG-ODN) on the immune response to an antigen applied to tape-stripped mouse skin, by evaluating the production of cytokines and Ig isotypes. Confocal laser scanning microscopy revealed that the Ovalbumin (OVA) (model antigen) and CpG-ODN easily penetrated the tape-stripped skin. Coadministration of CpG-ODN and OVA to the disrupted skin elicited an antigen-specific, Th1-predominant immune response, and enhanced the production of Th1-type cytokines, IL-12, and IFN-γ (Interferon). On the other hand, the production of a Th2-type cytokine, IL-4, was drastically suppressed. In terms of antigen-specific antibody production, the IgG2a level, which is regulated by IFN-γ, was increased by CpG-ODN, but IgE production regulated by IL-4 was suppressed. Furthermore, the administration of CpG-ODN through the skin drastically attenuated the production of IgE in mice experiencing IgE-type immune response. The administration of CpG-ODN through the skin may shift the immune response from a Th2 to a Th1-like response.

Continuing with their studies, Inoue et al. (2007) also demonstrated that TS induces the expression of toll-like receptor (TLR)-9 in the skin, and enhances the Th1-type immune response triggered by CpG-ODN administered through the tape-stripped skin. TS induces the expression of TLR-9 and tumor necrosis factor (TNF)-α in the skin, and CpG-ODN treatment through the tape-stripped skin enhances the migration of APCs to the draining lymph nodes. On the other hand, TLR-9 mRNA and TNF-α mRNA were not observed in the skin when CpG-ODN was injected intradermally, or in Th1-type immune response. The transdermal application of CpG-ODN with an antigen through the tape-stripped skin is an effective way to induce a Th1-type immune response and is also a simple, cost-effective, and needle-free vaccination system.

Liu et al. (2001) administered human immunodeficiency virus type-1 (HIV-1) DNA vaccine with cytokine-expressing plasmids to the skin of mice by a new topical application technique involving prior elimination of keratinocytes using TS. Their results revealed that the topical application of HIV-1 DNA vaccine induced high levels of both humoral and cell-mediated immune activity against HIV-1 envelope antigen. Coadministration of the DNA vaccine with cytokine expression plasmids of IL-12 and granulocyte-macrophage colony-stimulating factor by this new method raised the levels of both the HIV-specific cytotoxic T lymphocyte response and delayed-type hypersensitivity and facilitated the induction of substantial immune responses by DNA vaccine. Skin biopsy sections showed significant increases of S-100 protein-positive DCs. These results suggest that the topical application method is an efficient route of DNA vaccine administration and that the immune response may be induced by DNA plasmids taken in by DCs, LCs, or other such APCs. This new topical application is likely to be of benefit in clinical use.

4.3.3.1.13 Other Kind of Permeants (Polyethylene glycols, 4-Cyanophenol, Urea, Alkanols)

Ayala-Bravo et al. (2003) investigated the effect of sucrose esters (sucrose oleate and sucrose laureate in water or Transcutol, TC) on the SC barrier properties in vivo, and examined the impact of these surfactant-like molecules on the in vivo percutaneous penetration of a model penetrant, 4-hydroxybenzonitrile (4-cyanophenol, 4CP). The effect of the enhancers on 4CP penetration was monitored in vivo using ATR-FTIR spectroscopy in conjunction with TS of the treated site. A combination of sucrose esters (oleate or laureate) and TC is able to temporarily alter the SC barrier properties, thereby promoting 4CP penetration.

Results from TS experiments can be affected significantly by chemical diffusion into the SC during the time required to apply and remove all of the TSs, t_{TS} (period of time required to completely remove the SC by TS). For this reason, Reddy et al. (2002) studied the dermal absorption of 4CP in humans using TS experiments to assess the conditions under which diffusion alters TS results. Mathematical models were developed to assess the effects of diffusion on parameter estimation. In an experiment with $t_{TS} > t_{lag}$ (i.e., the lag time for a chemical to cross the SC), the permeability coefficient for 4CP, Psc,v, calculated including t_{TS}, was consistent with the values from the literature. When diffusion during stripping was not included in the model, Psc,v, was 70% smaller. Calculations show that chemical concentrations in TSs can be affected by diffusion during TS, but with $t_{TS} < 0.2\ t_{lag}$ and an exposure time $> 0.3\ t_{lag}$, TS concentrations are not significantly affected by t_{TS}.

Jakasa et al. (2004) developed a sensitive method for the determination of polyethylene glycols (PEGs) with different molecular weights (MW) in the human SC obtained by TS. The analysis is based on derivatization with pentafluoropropionic anhydride and gas chromatography–electron capture detection. The method showed to be suitable for studying permeability in normal and impaired skin with respect to MW in the range of 150–600 Da.

In order to obtain more data to assess the barrier function of uninvolved skin in atopic dermatitis (AD) patients, Jakasa et al. (2007) determined the percutaneous penetration of PEGs of various molecular sizes in vivo in AD patients and control subjects using TS of the SC. The apparent diffusion coefficient of PEGs through atopic skin was twice as high as that through normal skin, and decreased with increasing MW in both groups. The partition coefficient in the skin of AD patients was half of that for normal skin, but as for normal skin, there was no MW dependence. Although atopic skin exhibited an altered barrier with respect to diffusion and partitioning, the permeability coefficients were nearly the same for atopic and normal skin. The results support the assumption of an altered skin barrier in AD patients, even if the skin is visibly unaffected by the disease.

Tsai et al. (2003) further investigated the dependence of permeability on MW with different forms of barrier disruption. A series of PEGs with a MW ranging from nearly 300 to over 1000 Da were used to study the effects of TS and sodium dodecyl sulfate (SDS) treatment on MW permeability profiles of mouse skin in vitro. The total penetration of PEG oligomers across control skin and tape-stripped skin and SDS treated to different degrees of barrier disruption progressively decreased with increasing MW. Penetration enhancement relative to control skin was more prominent with larger molecules. The MW cutoff for skin penetration increased with the degree of barrier disruption, irrespective of the treatment applied, and was 986 Da (TS) and 766 Da (SDS treatment) at TEWL levels in the range of 10–20 g/m^2 per h, compared with 414 Da for control skin. The results strongly suggest that, regardless of the form of barrier disruption applied, not only higher amounts, but also more varieties of chemicals (larger molecules), may penetrate into the skin in the presence of a compromised barrier compared with normal skin.

Park et al. (1995) measured the status of skin surface hydration of uraemic patients with a corneometer and skin surface hydrometer, the functional capacity and the urea concentration of SC by TS technique, and the response of eccrine sweat gland to sudorific agent (0.05% pilocarpine HCL, hydrochloride) in 18 age-matched hemodialysis patients and 10 healthy volunteers. They also performed the water sorption–desorption test to uraemic and control subjects after application of urea in various concentrations. Uraemic patient's skin showed decreased water content compared to control subjects. However, they found no correlation between dry skin and pruritus. Although the urea concentration determined by TS of the horny layer in uraemic patients was elevated compared to control subjects (28.2 vs. 5.04 µg/cm^2), its moisturizing effect to relieve pruritus is questionable because its artificial application revealed no improvement of the functional capacity of the horny layer in concentration five times higher than the physiological concentration. Uraemic patients showed a decreased sweating response to the sudorific agent. In conclusion, the functional abnormalities of eccrine sweat glands may be accounted for by dry skin in uraemic patients at least in part, but there is no correlation between xerosis and pruritus.

Dias et al. (2008) studied the higher alkanols (hexanol, octanol, and decanol) in vivo using a combination of ATR-FTIR spectroscopy and TS. Studies conducted in vivo using deuterated vehicles confirmed the lipid extraction effects of D-hexanol and D-octanol, whereas D-decanol did not change the skin lipid content. The uptake of D-decanol was higher than that for the other vehicles consistent with previous observations on mouse skin for alkanols of increasing chain length. In general, solvent uptake was proportional to the induced shift in the C–H stretching frequency. Lipid disorder was induced by all vehicles studied in vivo and was proportional to the amount of vehicle present in the skin.

4.3.3.1.14   UV Absorbers

Alvarez-Román et al. (2004) determined whether encapsulation of lipophilic compounds in polymeric nanoparticles is able to improve topical delivery to the skin. The penetration of octyl methoxycinnamate (OMC) encapsulated in poly(ε-caprolactone) nanoparticles, into and across porcine ear skin in vitro, was investigated using TS.

The quantification of OMC in the skin using TS demonstrated that nanoparticulate encapsulation produced a 3.4-fold increase in the level of OMC within the SC. The nanoparticulate encapsulation of OMC increased its "availability" within the SC.

Olvera-Martínez et al. (2004) prepared polymeric NCs containing OMC, and their in vivo distribution profile through the SC was determined by the TS technique. The penetration degree of OMC formulated in NCs was compared with that obtained for a nanoemulsion (NE) and a conventional oil-in-water (o/w) emulsion (EM). In vivo percutaneous penetration, evaluated by the TS technique, demonstrated that NE increased the extent of OMC penetration relative to the penetration achieved by NCs or EM. Likewise, OMC accumulation in the skin was significantly greater with NE than with EM or NCs.

Sarveiya et al. (2004) developed a reverse HPLC assay to quantify four common sunscreen agents, namely, 2-hydroxy-4-methoxybenzophenone, 2-ethylhexyl-*p*-methoxycinnamate, octylsalicylate, and SA 3,3,5-trimethcyclohexyl ester in a range of biological matrices. This assay was further applied to study skin penetration and systemic absorption of sunscreen filters after topical application to human volunteers. The assay allows the analysis of sunscreen agents in biological fluids, including bovine serum albumin solution, plasma, and urine, and in human epidermis by using the TS technique. The results from the preliminary clinical study demonstrate a significant penetration of all sunscreen agents into the skin.

Lademann et al. (2004b) determined the amount of sunscreen present on the skin of people at the beach. The amounts of sunscreen applied to different body sites were quantitatively determined by TS. The actual amounts of sunscreen applied were compared with the European Cosmetic Toiletry and Perfumery Association (COLIPA) standard. Most volunteers had applied 10% or less of the COLIPA standard amount to all body sites assessed.

Mavon et al. (2007) assessed the penetration of titanium dioxide (TiO_2) and methylene bis-benzotriazoyl tetramethylbutylphenol (MBBT), included in a broad-spectrum sunscreen formulation, into human skin in vivo, using the TS technique, and in vitro, using a compartmental approach. More than 90% of both sunscreens were recovered in the first 15 TSs. In addition, they have shown that the remaining 10% did not penetrate the viable tissue, but was localized in the furrows. Less than 0.1% of MBBT was detected in the receptor medium, and no TiO_2 was detected in the follicles, the viable epidermis, or the dermis. Thus, this in vivo and in vitro penetration study showed an absence of TiO_2 penetration into viable skin layers through either transcorneal or transfollicular pathways, and a negligible transcutaneous absorption of MBBT. However, differences in distribution within the SC reinforced the need for a complementary approach, using a minimally invasive in vivo methodology and an in vitro compartmental analysis. This combination represents a well-adapted method for testing the safety of topically applied sunscreen formulations in real-life conditions.

4.3.3.1.15 Fragrances

In vitro human skin permeation and distribution of geranyl nitrile (GN) were determined by Brian et al. (2007) using epidermal membranes, following application in 70% ethanol, under non-occlusive conditions, at maximum in-use concentration (1%).

Levels of GN in the epidermis (plus any remaining in SC after TS), filter paper membrane support, and receptor fluid were combined to produce a total absorbed dose value of 4.72 ± 0.32%. The systemic exposure resulting from the use of GN as a fragrance ingredient, under unoccluded conditions, would be low based on the currently reported use levels.

4.3.3.1.16 Dyes

Teichmann et al. (2006) developed a method to investigate the effectiveness of reservoir closure by different formulations. Patent Blue V in water was used as a model penetrant. Its penetration, with and without barrier cream treatment, was analyzed by TS in combination with UV–VIS spectroscopic measurements. The investigations showed that the SC represents a reservoir for topically applied Patent Blue V in water. Furthermore, the barrier investigations showed that Vaseline and bees wax form a 100% barrier on the skin surface. The third barrier cream, containing waxes and surfactant, only partially showed a protective effect against the penetration of Patent Blue V in water. Strong inter-individual differences were observed for this barrier product. In conclusion, it was assumed that the application of barrier creams cannot replace other protective measures, and should be used to inhibit low-grade irritants or in combination with other protectants, or in body areas where other protective measures are not applicable.

Jacobi et al. (2006) studied the penetration of highly hydrophilic (Patent Blue V) and lipophilic (curcumin) dyes into the skin using pure oil (o) or water (w), and comparing them with an o/w EM. The penetration and localization of both dyes were investigated in vivo using TS and microscopy techniques. Differences in the distribution and the localization of both dyes within the SC were observed. These differences depend on the physicochemical properties of both the vehicles and the dyes. The vehicle appears to affect, in particular, the penetration pathways.

Coderch et al. (1999) studied the reservoir capacity of the SC by topical application of sodium fluorescein encapsulated in vesicles in order to elucidate the mechanism involved in the (trans)dermal transport of drugs when vesicles are applied to the skin. The penetration profile of sodium fluorescein in the different strips was found to be logarithmic with the constant and the slope of the regression curves accounting for the superficial non-penetration content and for the penetration rate inside the SC, respectively. The results show a small but significantly enhanced penetration of these vesicle structures for the release of hydrophilic substances. Moreover, similar values obtained when the same liposomes with varying encapsulation content were applied could lend support to the penetration mechanism where the vesicle enhancement is mainly due to SC structural modification.

As we can observe, there is an ongoing search for the identification of testing methods to optimize topical dosage forms and to assess topical drug BA. While in vitro screening continues to play an important role (and is relatively inexpensive and easy to use), regulatory approval of drug delivery systems to the skin, with few exceptions, requires clinical trials to be performed. For many drugs used topically, the problem remains unsolved, since an easily visualized pharmacodynamic response is not elicited.

As a consequence, various alternative techniques have been considered, of which SC TS is being given the greatest attention (Escobar-Chávez et al. 2008, 2009). While the former is technically more challenging, the potential reward is a drug concentration–time profile in a compartment presumed to be in close communication with the site of action of most dermatological drugs. The latter, in contrast, offers an apparently easy and quite noninvasive methodology for skin tissue sampling, and is the basis of the FDA's so-called DPK. Validation and optimization of the procedure have not come quickly. The goal of the research described here is not only to contribute to further establishing the credibility of the TS technique, but also to demonstrate that useful and relevant measurements can be made on a surrogate, ex vivo skin model.

4.3.3.2 Dermal Absorption of Toxic or Irritant Chemicals

The rate and extent of dermal absorption are important in the analysis of risk from dermal exposure to toxic chemicals, and for the development of topically applied drugs, barriers, insect repellents, and cosmetics, and the TS technique has been widely used to determine the penetration of these kinds of substances (Nylander-French et al. 2000, Chao et al. 2004, Mattorano et al. 2004, de Jongh et al. 2006, Lundgren et al. 2006, Van der Merwe et al. 2006).

Mattorano et al. (2004) developed and tested a simple, noninvasive dermal sampling technique on 22 human volunteers to estimate acute dermal exposure to jet fuel (JP-8). Two sites on the ventral surface of each forearm were exposed to 25 µL of JP-8, and the SC was sequentially tape stripped using an adhesive tape. The analysis of the first tape strip indicated that JP-8 was rapidly removed from the SC over the 20 min study period. On average, after 5 min of exposure, the first two tape strips removed 69.8% of the applied dose. The amount recovered with two tape strips decreased over time, to a recovery of 0.9% 20 min after exposure. By fitting a mixed-effect linear regression model to the TS data, the authors were able to accurately estimate the amount of JP-8 initially applied. This study indicates that naphthalene has a short retention time in the human SC and that the TS technique, if used within 20 min of initial exposure, can be used to reliably measure the amount of naphthalene initially present in the SC due to a single exposure to jet fuel.

Chao et al. (2004) presented a TS method for the removal and quantification of keratin from the SC for normalization of extracted concentrations of naphthalene (as a marker for jet fuel exposure) from 12 human volunteers before and after exposure to jet fuel (JP-8). Due to the potential for removal of variable amounts of squamous tissue from each tape strip sample, keratin was extracted and quantified using a modified Bradford method. Naphthalene was quantified in the sequential tape strips collected from the skin between 10 and 25 min after a single dose of JP-8 was initially applied. The penetration of jet fuel into the SC was demonstrated by the fact that the average mass of naphthalene recovered by a tape strip decreased with increasing exposure time and subsequent tape strips. The actual concentration of naphthalene (as a marker for JP-8 exposure) per unit of keratin in a tape strip sample can be determined by using this method, and may prove necessary when measuring occupational exposures under field conditions.

Van der Merwe et al. (2006) described a physiologically based pharmacokinetic (PBPK) model developed to simulate the absorption of organophosphate pesticides, such as parathion, fenthion, and methyl parathion through porcine skin with flow-through cells. Three parameters were optimized based on experimental dermal absorption data, including solvent evaporation rate, diffusivity, and a mass transfer factor. Diffusion cell studies were conducted to validate the model under a variety of conditions, including different dose ranges (6.3–106.9 µg/cm^2 for parathion; 0.8–23.6 µg/cm^2 for fenthion; and 1.6–39.3 µg/cm^2 for methyl parathion), different solvents (ethanol, 2-propanol, and acetone), different solvent volumes (5–120 µL for ethanol; 20–80 µL for 2-propanol and acetone), occlusion versus open to atmosphere dosing, and corneocyte removal by TS. The study demonstrated the utility of PBPK models for studying dermal absorption. The similarity between the overall shapes of the experimental and model-predicted flux/time curves and the successful simulation of altered system conditions for this series of small, lipophilic compounds indicated that the absorption processes described in the model successfully simulated important aspects of dermal absorption in flow-through cells. These data have a direct relevance in the assessment of topical organophosphate pesticides' risk.

De Jongh et al. (2006) studied whether sodium lauryl sulfate (SLS) penetration rate into the SC is related to an impairment of skin's water barrier function and inflammation. The penetration of SLS into the SC was assessed using a noninvasive TS procedure in 20 volunteers after a 4 h exposure to 1% SLS. Additionally, the effect of a 24 h exposure to 1% SLS on the skin water barrier function was assessed by measuring TEWL. A multiple regression analysis showed that the baseline TEWL, SC thickness, and SLS penetration parameters K (SC/water partition coefficient) and D clearly influenced the increase in TEWL after the 24 h irritation test. They found that variation in barrier impairment and inflammation of human skin depends on SLS' penetration rate, which was mainly determined by SC thickness.

Lundgren et al. (2006) modified and tested a vacuuming sampler for removing particles from the skin. The sampler was compared with two other skin and surface exposure sampling techniques. These were based on surrogate skin (a patch sampler-adhesive tape on an optical cover glass) and a TS removal procedure. All three samplers measure the mass of dust on the skin. Dust containing starch was deposited onto the skin in a whole-body exposure chamber. Samples were taken from forearms and shoulders and analyzed using optical microscopy. With the different sampling techniques, small differences in results were obtained. There was a good agreement between the vacuuming sampler and the TS technique.

Nylander-French et al. (2000) developed and tested a noninvasive sampling method for measuring dermal exposure to a multifunctional acrylate employing a TS technique of the SC. Samples were subsequently extracted and a gas chromatographic method was employed for quantitative analysis of tripropylene glycol diacrylate (TPGDA). This method was tested in 10 human volunteers exposed to an a priori determined amount of TPGDA or a UV-radiation curable acrylate coating containing TPGDA (UV-resin) at different sites on hands and arms. On the average, the first TS removed 94% of the theoretical quantity of deposited TPGDA and 89% of the theoretical quantity of deposited TPGDA in UV-resin 30 min after exposure. Quantities of TPGDA recovered from two consecutive TSs accounted for the entire test agent, demonstrating both the efficiency of the method to measure dermal exposure and the potential to determine the rate of absorption with successive samples over time. In general, the amount removed by the first TS was greater for TPGDA than for UV-resin while the second TS removed approximately 6% and 21% of TPGDA and UV-resin, respectively. However, when the amounts removed with the first TS for TPGDA or UV-resin from the five different individual sites were compared, no significant differences were observed. No significant difference was observed in recovery between TPGDA and UV-resin for the first TS when calculated as a percentage of the theoretical amount. The results indicate that this TS technique can be used to quantify dermal exposure to multifunctional acrylates.

4.4 OTHER APPROACHES TO QUANTIFY DRUGS THROUGHOUT THE SKIN

Other techniques that can be mentioned are the suction blister technique and skin biopsy (either with a blade, a punch, or with the use of cements or glues) (Darlenski et al. 2008, Tettey-Amlalo et al. 2009). Comparative characteristics of skin biopsy and skin FBS techniques for the clinical assessment of tissue drug distribution in humans are shown in Table 4.2.

4.4.1 SUCTION BLISTER SAMPLING TECHNIQUE

The suction blister technique involves the application of vacuum (about 150–250 mm Hg) during a variable time (30 min to 2–3 h) by means of commercially available devices (Svedman, 1996, Kaliyadan et al. 2008). These devices consist of a suction

TABLE 4.2
Comparative Characteristics of Skin Biopsy and Skin Fluid Blister Sampling Techniques for the Clinical Assessment of Tissue Drug Distribution in Humans

Characteristics	Skin Biopsy	Skin Blister Fluid
Matrix sampled measured	Total tissue	Blister fluid
Invasiveness	Invasive	Semi invasive
Direct measurement of unbound drug concentrations	No	No
Continuous monitoring	No	No
Technical complexity	Low	Low
Cost	Low	Low
Major disadvantages affecting the use in clinical drug distribution studies	Invasiveness; hybrid tissue concentration owing to mixture of different compartments	Difficult to standardize; contains proteins; limited number of time points

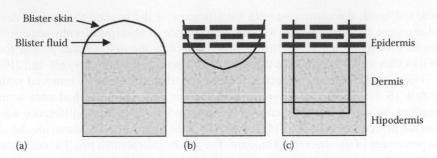

FIGURE 4.5 Schematic representation of skin sampling techniques. (a) Suction blister, (b) shave biopsy, and (c) punch biopsy.

cup with holes. The hole size can vary from 3 to 10 mm and the total diameter of the cup from 20 to 50 mm. A blister is formed as a product of the separation epidermis–dermis, at the level of lamina lucida (Figure 4.5a). The blister cavity formed is filled with tissue fluid or interstitial fluid (Herfst and van Rees, 1978, Vermeer et al. 1979; Vessby et al. 1987). Then, the transepidermal absorption of a drug can be studied by sampling the blister fluid at various time intervals.

The suction blister model was originally developed in 1968 by Kiistala et al., for the separation of viable epidermis from dermis; it has shown to be useful to perform pharmacokinetic and pharmacodynamic evaluations. In vivo sampling may be done after topical or systemic administration of the drug (Groth and Staberg, 1984, Mouton et al. 1990, Vaillant et al. 1991, Surber et al. 1993, Mazzei et al. 1994, Benfeldt et al. 1999, Toyoda et al. 2007). A close correlation for the pharmacokinetic parameters such as Cmax (maximum concentration) and Tmax (time to Cmax) has been found between blister fluid and plasma. Since its development, its use has been expanded and applied to several other applications. One new application assesses the collagen synthesis rate in the human skin in vivo by using the suction blister fluid collected in an assay of collagen propeptides (Oikarinen et al. 1982, 1992, Ihlberg et al. 1993, Autio et al. 1996). This method is sensitive and can detect changes in collagen synthesis owing to various diseases and topical or systemic therapies. Other applications of the suction blister technique include measurements of pharmacological agents or their derivatives from interstitial fluid, and assays of various enzymes, cytokines, and so on (Vessby et al. 1987, Punnonen et al. 1991, Haaverstad et al. 1996, Leivo et al. 2000). Additionally, the suction blister technique has even been used to treat vitiligo by collecting living melanocytes from blister roofs from healthy skin and injecting these into blister cavities induced in diseased skin.

The suction blister technique separates epidermis from dermis within the basement membrane (Kiistala 1992), this model can also be used to study reepithelization and skin barrier regeneration of superficial wounds of a standard size. The level of separation of epidermis from dermis occurs above type IV collagen, since this collagen remains completely in the blister floor after blistering (Kiistala et al. 1984).

4.4.2 Use of Suction Fluid Blister Sampling Technique to Quantify the Penetration of Drugs

Applying a partial negative pressure to the skin disrupts the epidermal–dermal junction and forms a blister, which fills progressively with interstitial fluid and serum (Volden et al. 1980, Svedman and Svedman 1998). This liquid offers a pharmacokinetic compartment, therefore, in which a previously applied drug can be sampled with a hypodermic needle and quantified; if multiple blisters are raised, then a concentration–time profile of the drug in the skin can be obtained. However, this approach is also quite invasive and causes obvious scarring, albeit over relatively small areas of skin. The technique can be used to compare topical formulations in a reasonably objective way (Makki et al. 1991, Treffel et al. 1991), but the potential binding of the drug to skin tissue, especially for highly lipophilic species, may mean that very low levels are present, if at all, in the blister fluid (Surber et al. 1993). Taken together, this has meant that the procedure has not been widely used, and, once again, is presently viewed as too invasive for practical and routine application in topical BA experiments. Table 4.3 summarizes the research with FBS technique to determine the penetration and kinetics of some drugs.

4.4.2.1 Analgesics

Benfeldt et al. (1999) simultaneously investigated two techniques for in vivo sampling of peripheral compartment pharmacokinetics after systemic administration of acetylsalicylic acid. Ten volunteers were given 2 g acetylsalicylic acid orally. Blood samples and dialysates from four microdialysis probes inserted in the dermis of the forearm were collected for 5 h and suction blisters were raised 1–3 h after dosing. In microdialysates, both acetylsalicylic acid and the metabolite SA were measurable in the absence of hydrolysing enzymes. The mean C_{max} of total, unbound SA was 9.5 μg/mL in microdialysates, 13.2 μg/mL in suction blister fluid, and 56.5 μg/mL in plasma. Mean T_{max} for SA was 188 and 161 min in plasma and microdialysates, respectively. The dermis-to-plasma C_{max} ratio was 0.16 ± 0.04 by microdialysis sampling and 0.25 ± 0.09 by the suction blister fluid method. Close correlations were found between C_{max} of SA in microdialysates and plasma, and between C_{max} of SA in suction blister fluid and plasma. The two techniques were in excellent accordance with even a closer correlation between maximum concentrations obtained by microdialysis and suction blister fluid sampling. However, comparing the tolerability of the sampling procedure, ease of analysis, and detail in chronology, microdialysis is superior for sampling in vivo pharmacokinetics in the dermis.

Walker et al. (1993) after a single oral dose of (R,S)-ibuprofen (1200 mg) to five healthy volunteers, paired plasma and blister fluid concentrations of drug were determined by a stereospecific HPLC assay. A pharmacokinetic model that incorporated blister fluid as a separate peripheral compartment adequately characterized the data. The plasma concentrations were consistently higher for (S)-ibuprofen than (R)-ibuprofen in both plasma and blister fluid. No significant difference in the elimination half-life of the enantiomers was observed. Similar to synovial fluid, there were relatively small fluctuations in blister fluid concentrations of both enantiomers.

TABLE 4.3
Research on the Fluid Blister Sampling Technique as a Method to Determine Skin Penetration of Different Kind of Permeants

Research	Outcome	Author Year
Kinetics and Penetration Depth of Drugs		
Analgesic and Anti-Inflammatory Drugs		
Investigation of two techniques for in vivo sampling of peripheral compartment pharmacokinetics after systemic administration of acetylsalicylic acid.	The two techniques were in excellent accordance with even closer correlation between maximum concentrations obtained by microdialysis and suction blister fluid sampling. However, comparing the tolerability of the sampling procedure, ease of analysis, and detail in chronology, microdialysis is superior for sampling in vivo pharmacokinetics in the dermis.	Benfeldt et al. (1999)
Paired plasma and blister fluid concentrations of drug determined by a stereospecific HPLC assay after a single oral dose of (R,S)-ibuprofen (1200 mg) to five healthy volunteers,	This use of skin blisters, which can be sampled repetitively, may therefore prove to be a valuable experimental technique in pharmacokinetic and pharmacodynamic studies of drugs, especially in patients in whom synovial fluid is not available for sampling.	Walker et al. (1993)
Antibiotics		
To characterize the penetration of moxifloxacin into peripheral target sites, by measuring unbound moxifloxacin concentrations in the interstitial space fluid by means of microdialysis.	Moxifloxacin concentrations attained in the interstitial space fluid in humans and in skin blister fluid following single doses of 400 mg exceed the values for the MIC at which 90% of isolates are inhibited for most clinically relevant bacterial strains, notably including penicillin-resistant *Streptococcus pneumoniae*. These findings support the use of moxifloxacin for the treatment of soft tissue and respiratory tract infections in humans.	Müller et al. (1999)
Antimicotics		
The uptake of fluconazole into the interstitial fluid of human subcutaneous tissue using the microdialysis and suction blister techniques.	There was good agreement between fluconazole concentrations derived from microdialysis sampling and those estimated using a blood-flow-limited pharmacokinetic model.	Sasongko et al. (2002)
Antiviral Drugs		
Comparison of the suction blister technique and microdialysis as methods to measure the penciclovir concentration in the skin after a single dose (250 mg) of its prodrug, famciclovir.	Microdialysis and the suction blister technique can be used to study the time–concentration profile of penciclovir in the skin and microdialysis allows a continuous sampling of the drug for a prolonged time after administration.	Borg et al. (1999)

TABLE 4.3 (continued)
Research on the Fluid Blister Sampling Technique as a Method to Determine Skin Penetration of Different Kind of Permeants

Research	Outcome	Author Year
	Theophillyne	
To measure the tissue kinetics of theophylline following single dose administration simultaneously in cantharides induced skin blisters, saliva, and microdialysates of subcutaneous- and skeletal muscle tissue and compared to plasma concentrations.	It was concluded that microdialysis represents a reliable technique for the measurement of unbound peripheral compartment concentrations and is superior to saliva and skin blister concentration.	Müller et al. (1998)
	Sulphas	
Determination of the penetration of trimethoprim, sulphamethoxazole and its main metabolite–N4-acetylsulphamethoxazole into cantharidin-induced skin blister fluid following administration of a single oral combination dose of 320 mg trimethoprim and 1600 mg sulphamethoxazole.	The cantharidin-induced skin blister is a useful technique to determine the penetration into skin of a drug and its metabolite and to evaluate pharmacokinetic parameters.	Królicki et al. (2002)
	Proteins	
To evaluate whether the simpler blister suction technique using large blisters could replace the wick technique in the investigation of patients with postreconstructive leg edema.	There was a good methodological agreement between the blister suction technique and the wick technique. The less invasive blister suction technique should be regarded as the method of choice for the investigation of subcutaneous interstitial tissue fluid in patients with postreconstructive leg edema.	Haaverstad et al. (1996)
	Zinc	
Penetration of zinc through normal skin treated with a zinc oxide (25%) in medicated occlusive dressing.	The study shows that zinc permeates intact human skin from topically applied zinc oxide in vivo.	Agren (1990)

Blister fluid, similar to synovial fluid, therefore behaves pharmacokinetically as a peripheral compartment for drug distribution. This use of skin blisters, which can be sampled repetitively, may therefore prove to be a valuable experimental technique in pharmacokinetic and pharmacodynamic studies of drugs, especially in patients in whom synovial fluid is not available for sampling.

4.4.2.2 Antibiotics

To characterize the penetration of moxifloxacin into peripheral target sites, Müller et al. (1999) measured unbound moxifloxacin concentrations in the interstitial space fluid by means of microdialysis, an innovative clinical sampling technique. In addition, moxifloxacin concentrations were measured in cantharides-induced skin blisters, saliva, and capillary plasma and compared to total- and free-drug concentrations in venous plasma. For this purpose, 12 healthy volunteers received moxifloxacin in an open randomized crossover fashion either as a single oral dose of 400 mg or as a single intravenous infusion of 400 mg over 60 min. An almost-complete equilibration of the free unbound plasma fraction of moxifloxacin with the interstitial space fluid was observed, with mean area under the concentration–time curve $AUC_{interstitial\ fluid}$/ $AUC_{total\text{-}plasma}$ ratios ranging from 0.38 to 0.55 and mean $AUC_{interstitial\ fluid}$/$AUC_{free\text{-}plasma}$ ratios ranging from 0.81 to 0.86. The skin blister concentration/plasma concentration ratio reached values above 1.5 after 24 h, indicating a preferential penetration of moxifloxacin into inflamed lesions. The moxifloxacin concentrations in saliva and capillary blood were similar to the corresponding levels in plasma. Their data show that moxifloxacin concentrations attained in the interstitial space fluid in humans and in skin blister fluid following single doses of 400 mg exceed the values for the MIC at which 90% of isolates are inhibited for most clinically relevant bacterial strains, notably including penicillin-resistant *Streptococcus pneumoniae*. These findings support the use of moxifloxacin for the treatment of soft tissue and respiratory tract infections in humans.

4.4.2.3 Antimicotics

Sasongko et al. (2002) investigated the uptake of fluconazole into the interstitial fluid of human subcutaneous tissue using microdialysis and suction blister techniques.

A sterile microdialysis probe (CMA/60) was inserted subcutaneously into the upper arm of five healthy volunteers following an overnight fast. Blisters were induced on the lower arm using gentle suction prior to ingestion of a single oral dose of fluconazole (200 mg). Microdialysate, blister fluid, and blood were sampled over 8 h. Fluconazole concentrations were determined in each sample using a HPLC assay.

In vivo recovery of fluconazole from the microdialysis probe was determined in each subject by perfusing the probe with fluconazole solution at the end of the 8 h sampling period. Individual in vivo recovery was used to calculate fluconazole concentrations in subcutaneous interstitial fluid. A PBPK model was used to predict fluconazole concentrations in human subcutaneous interstitial fluid.

There was a lag-time (approximately 0.5 h) between detection of fluconazole in microdialysate compared with plasma in each subject. The in vivo recovery of fluconazole from the microdialysis probe ranged from 57.0% to 67.2%. The subcutaneous interstitial fluid concentrations obtained by microdialysis were very similar to the unbound concentrations of fluconazole in plasma with maximum concentration of 4.29 ± 1.19 mg/mL in subcutaneous interstitial fluid and 3.58 ± 0.14 mg/mL in plasma. Subcutaneous interstitial fluid-to-plasma partition coefficient of fluconazole was 1.16 ± 0.22. By contrast, fluconazole concentrations in blister fluid were significantly lower than unbound plasma concentrations over the first 3 h and maximum concentrations in blister fluid had not been achieved at the end of the sampling period. There was good agreement between fluconazole concentrations derived from microdialysis sampling and those estimated using a blood-flow-limited pharmacokinetic model.

4.4.2.4 Antiviral Drugs

Penciclovir is a drug active against herpes simplex viruses located in the epidermis basal layer. Borg et al. (1999) compared the suction blister technique and microdialysis as methods to measure the penciclovir concentration in the skin after a single dose (250 mg) of its prodrug, famciclovir. Suction blister fluid, microdialysates, and plasma were sampled from 11 healthy volunteers for 5 h after famciclovir administration. Both suction blister technique and microdialysis showed that penciclovir reaches the skin in concentrations sufficient to inhibit herpes virus replication. The maximum concentration in both suction blister fluid and microdialysate was observed later than in plasma. The microdialysis concentration was decreased by cooling of the skin surface and by adrenaline-mediated vasoconstriction. The microdialysis recovery of penciclovir was studied with respect to the flow rate of perfusion medium through the microdialysis probe. Microdialysis and the suction blister technique can be used to study the time–concentration profile of penciclovir in the skin and microdialysis allows a continuous sampling of the drug for a prolonged time after administration.

4.4.2.5 Theophillyne

In many cases, the concentration reached in a peripheral effect compartment rather than in plasma determines the clinical outcome of the therapy. Therefore, several experimental approaches have been developed for direct assessment of drug kinetics in peripheral compartments.

Particularly, saliva sampling, shin blister fluid sampling, and in vivo microdialysis are frequently employed for measuring peripheral drug concentrations. The study by Müller et al. (1998) consisted of measuring the tissue kinetics of theophylline following single dose administration simultaneously in cantharides-induced skin blisters, saliva, and microdialysates of subcutaneous and skeletal muscle tissue and compared plasma concentrations. Theophylline was administered to nine healthy volunteers as an IV infusion of 240 mg. Mean ratio ($AUC_{saliva}/AUC_{plasma}$) was 0.63 ± 0.05, mean ratio ($AUC_{blister}/AUC_{plasma}$) was 0.69 ± 0.12, mean ratio ($AUC_{muscle}/AUC_{plasma}$) 0.41 + 0.10 and mean ratio ($AUC_{subcutaneous}/AUC_{plasma}$) was 0.34 ± 0.07. The time course of the concentration$_{peripheral}$/concentration$_{plasma}$ ratios showed that tissue concentrations obtained by microdialysis were closely correlated to free plasma levels, whereas saliva and cantharides blister data overestimated the corresponding free plasma concentrations. It was concluded that microdialysis represents a reliable technique for the measurement of unbound peripheral compartment concentrations and is superior to saliva and skin blister concentration.

4.4.2.6 Sulphas

Królicki et al. (2002) determined the penetration of trimethoprim, sulfamethoxazole, and its main metabolite—N4-acetylsulphamethoxazole—into cantharidin-induced skin blister fluid following administration of a single oral combination dose of 320 mg trimethoprim and 1600 mg sulphamethoxazole. Moreover, penetration of the two drugs into skin blister fluid was compared with penetration into the theoretical peripheral compartment calculated on the basis of plasma levels found. The material consisted of 12 male patients with bacterial skin diseases. Skin blisters were induced by applying 0.25% cantharidin ointment. Drug concentrations in plasma and skin blister fluid were measured with HPLC. Peak concentrations of trimethoprim in plasma and skin blister fluid were $8.5 \pm 1.1 \mu mol/L$ after 3 h and $5.6 \pm 0.8 \mu mol/L$ after 7 h. In the theoretical peripheral compartment, peak concentration was $5.8 \pm 2.2 \mu mol/L$ after 9 h. Half-times of trimethoprim in plasma and skin blister fluid were 11.1 ± 4.5 and 12.3 ± 4.9 h, respectively. The degree of drug penetration into blister fluid defined as the ratio of area under concentration–time curves for blister fluid and plasma was 0.94 ± 0.23. Peak concentrations of sulphamethoxazole in plasma and skin blister fluid were $295 \pm 47 \mu mol/L$ after 3 h and $182 \pm 46 \mu mol/L$ after 8 h. The differences between both compartments as to parameters measured were statistically significant. In the theoretical peripheral compartment, peak concentration was $239 \pm 58 \mu mol/L$ after 7 h. Half-times of sulphamethoxazole in skin blister fluid and plasma were 9.7 ± 3.3 and 10.0 ± 1.1 h, respectively, and did not differ significantly. The drug penetrated into the blister fluid to a high extent, although less than trimethoprim, the degree of penetration being 0.82 ± 0.21. Peak concentration of N-4-acetylsulphamethoxazole, the main metabolite of sulphamethoxazole, was significantly lower in blister fluid than in plasma and took longer to achieve. The ratio of area under concentration–time curves in these two biological fluids of 0.86 ± 0.18 was similar to that of the parent drug. The results show that trimethoprim and sulphamethoxazole administered together penetrate from plasma into skin blister fluid to a great extent and achieve concentrations exceeding the MIC for susceptible pathogens. This finding confirms the usefulness of this treatment in bacterial skin diseases. The cantharidin-induced skin blister is a useful technique to determine the penetration into skin of a drug and its metabolite and to evaluate pharmacokinetic parameters.

4.4.2.7 Proteins

The wick technique and the blister suction technique are the most common methods for sampling of subcutaneous interstitial tissue fluid in man. The blister suction technique has the advantage of being less invasive than the wick technique, but the reliability of this method is still controversial. The aim of the study of Haaverstad et al. (1996) was to evaluate whether the simpler blister suction technique using large blisters could replace the wick technique in the investigation of patients with postreconstructive leg edema. Fifteen patients with ipsilateral leg edema following infrainguinal bypass surgery for lower limb atherosclerosis were investigated. The two different fluid sampling techniques were applied simultaneously on both legs. The concentration of total protein and albumin as well as the colloid osmotic pressure of the subcutaneous interstitial tissue fluid in the leg were measured in all fluid samples. The agreement index was found to be positive for all variables from the operated as well as from the contralateral control limb. Furthermore, all values were within the agreement limit. The best agreement between the two methods was found for colloid osmotic pressure on the operated side. According to the equation of linear regression, there was a slight overestimation of the wick values compared to the observed blister values. In conclusion, there was a good methodological agreement between the blister suction technique and the wick technique. The less invasive blister suction technique should be regarded as the method of choice for the investigation of subcutaneous interstitial tissue fluid in patients with postreconstructive leg edema.

4.4.2.8 Zinc Oxide

The penetration of zinc through normal skin treated with a zinc oxide (25%)-medicated occlusive dressing was studied by Agren (1990). The mean release rate of zinc to the skin was 5 µg/cm^2/h. After 48 h of treatment, suction blisters were raised by the "Kiistala method." The zinc concentration of the epidermis, blister fluid, and dermis was increased beneath the zinc dressing compared to control-treated skin. The study shows that zinc permeates intact human skin from topically applied zinc oxide in vivo.

4.4.3 Skin Biopsy

Skin biopsy consists in the direct excision of skin tissue. Skin biopsies, whether to the level of the dermis or through to the subcutis, are invasive and generally performed under local anesthesia, but it is an effective method to access skin compartments. If the SC is removed from the skin by TS, then the biopsy will contain epidermis, dermis, and subcutaneous tissue (Figure 4.5). Subcutaneous tissue can mechanically be separated, while epidermis and dermis can be separated by heating or by chemical or enzymatic techniques. Usually, a sample of 50 mg or less can be harvested.

Examination of a skin sample obtained by this method can be useful to diagnose a skin condition (e.g., psoriasis) and diseases of the skin (such as infections caused by bacteria or fungi, or even skin cancer).

There are different types of biopsies, including

- Excisional biopsy: An entire area of skin is cut out with a scalpel. Stitches are used to close the incision.
- Punch biopsy: A sharp cookie cutter–like instrument ranging in size from 1 to 8 mm is used to remove a small cylinder of skin (Figure 4.5c). Sometimes stitches are used to close the biopsy wound.
- Shave biopsy: A thin layer from the top of the skin is shaved with a scalpel (Figure 4.5b).
- Incisional biopsy: A small sample of a large lesion is removed with a scalpel. A punch biopsy is essentially an incisional biopsy, except that it is round. An advantage of the incisional biopsy over the punch method is that hemostasis can be done easier due to better visualization.
- Curettage biopsy: This can be done using a round curette blade.

To perform biopsy, the site must be cleansed and an anesthetic product (e.g., spray, cream, or injection) is applied. After biopsy, the patient may have some soreness around the site for 1 to 2 weeks, as well as a small scar. An analgesic may be recommended to relieve any discomfort.

Besides the use of biopsies to diagnose skin diseases, a biopsy can be very useful to determine depth of skin penetration of a drug, allowing comparisons between different formulations, administration routes, or conditions. As an example we can mention the work of Grundmann-Kollmann et al. (2002) who applied the punch biopsy method to study the distribution of 8-methoxypsoralen into the skin after systemic or topical application. Psoralens have shown to be effective in combination with photochemotherapy for the treatment of a variety of dermatoses.

This technique has also been applied to evaluate the effect of solvents on skin barrier function. In this sense, Rosado and Rodrigues (2003) use an in vivo innovative methodology, combining skin biopsy with colorimetry to characterize the action of different solvents commonly used as vehicles in transdermal systems, on skin barrier.

While such methods have a place in dermatological surgery (to remove warts and small tumors for example), their application for tissue sampling and analysis post drug application has not attracted much attention as a routine approach for use in vivo. Despite the obvious advantage of offering a snapshot of drug disposition in the different skin layers, the aggressive nature of the biopsy rules out any chance that it might be adopted as a standard procedure. Even attempts to minimize tissue trauma (Müller et al. 1998, Schrolnberger et al. 2001) fail to render the method remotely acceptable and its use in the foreseeable future will be restricted to animal and in vitro studies.

4.5 CONCLUSIONS

The quantification of drugs within the skin is essential for topical and transdermal delivery research. Despite pressure on regulatory agencies, such as the FDA, there is no generally accepted method with which to evaluate the BA and BE of topical drug products. For the present, therefore, with the exception of the vasoconstriction assay for corticosteroids, and despite the diversity of efforts that have been made, clinical studies are obligatory. Unlike oral administration, for example, where the blood level of a drug is a generally accepted surrogate for its concentration at the site of action, topical drug delivery poses a more complex problem.

Many in vivo methods for measuring dermal absorption of chemicals are invasive (e.g., blood sampling, skin biopsy and FBS) or slow (e.g., urine samples collected for extended periods). TS of the outermost skin layer, the SC, is a fast and relatively noninvasive technique for measuring the rate and extent of dermal absorption. The TS technique has the potential to meet the requirements for an efficient and reliable method to assess dermal exposures. TS has also been proposed as a method for evaluating the BE of topical dermatological dosage forms. DPK characterization of the penetration of active drugs in human volunteers has been suggested to be able to replace comparative clinical trials as a means of documenting BE. It is suggested that DPK assessment of drug concentrations in the SC is comparable to blood/urine measurements of systemically administered drugs, where the concentration of a drug in the SC is expected to relate to its concentrations in viable tissue. Short-contact DPK experiments can be used to obtain diffusion and partitioning parameters that may subsequently be able to predict drug penetration into the SC following longer application periods.

Although TS is widely used to determine dermal absorption through the SC, several factors can influence the actual technique. Recent reviews on this topic provide updated and additional insights (Choi et al. 2008, Escobar-Chávez et al. 2008, 2009, Lademann et al. 2008, Löffler et al. 2008). The investigation of variations in skin condition (dry vs. moist skin, skin defects, etc.) to determine their potential impact on the sampling method is warranted. For these reasons, the TS technique requires further development.

LIST OF ABBREVIATIONS

AD	Atopic dermatitis
APCs	Antigen-presenting cells
AQP3	Aquaporine-3
AsA	Ascorbic acid
AUC	Area under curve
BA	Bioavailavility
BE	Bioequivalence
BMV	Betamethasone 17 valerate
BPAA	4-biphenylylacetic acid
COLIPA	European Cosmetic Toiletry and Perfumery Association
CpG-ODN	CpG-oligodeoxynucleotide
4CP	4-Cyanophenol
CTL	Cytotoxic –T- lymphocyte
DCs	Dendritic cells
DM	Dexamethasone
DM-P	Dexamethasone phosphate
DMD	Dermal microdialysis
DPK	Dermatopharmacokinetic
EBA	Ethyl 4-biphenylyl acetate
EM	Emulsion
FDA	Food and Drug Administration
GN	Geranyl nitrile
HIV-1	Human immunodeficiency virus type-1
HP-P-CD	Hydroxypropyl-P-cyclodextrin
HPV	Human papilloma virus
IS	Impedance spectra
JP-8	Jet fuel-8
LCs	Langerhans cells
MBBT	Methylene bis-benzotriazoyl Tetramethylbutylphenol
MIC	Minimum inhibitory concentration
MO	Monoleine
MW	Molecular weight
NCs	Nanocapsules
NE	Nanoemulsion
OMC	Octyl methoxycinnamate
PG	Propylene glycol

RA   Retinyl chloride
RA-AsA   Retinyl ascorbate
SA   Salicylic acid
SC   Stratum corneum
SDS   Sodium dodecyl sulfate
SLs   Sesquiterpene lactones
SLS   Sodium lauryl sulfate
TA   Retinoic acid
TACA   Triamcinolone acetonide
TBF   Terbinafine
TC   Transcutol
TEWL   Transepidermal water loss
TLR-9   Toll-like receptor-9
TNF   Tumor necrosis factor
TPGDA   Tripropylene glycol diacrylate
TS   Tape stripping
TSs   Tape strippings
VZV   Varicella Zoster virus
WE   Water evaporation

ACKNOWLEDGMENTS

José Juan Escobar-Chávez would like to acknowledge the PROFIP/UNAM grant.

REFERENCES

Abdulmajed, K., Heard, Ch.M. 2004. Topical delivery of retinyl ascorbate co-drug 1. Synthesis, penetration into and permeation across human skin. *Int J Pharm.* 280:113–124.

Agren, M.S. 1990. Percutaneous absorption of zinc from zinc oxide applied topically to intact skin in man. *Dermatologica* 180(1):36–39.

Alberti, I., Kalia, Y.N., Naik, A., Guy, R.H. 2001. Assessment and prediction of the cutaneous bioavailability of topical terbinafine in vivo in man. *Pharm Res.* 18:1472–1475 (2001).

Alvarez-Román, R., Naik, A., Kalia, Y.N., Guy, R.H., Fessi, H. 2004. Enhancement of topical delivery from biodegradable nanoparticles. *Pharm Res.* 21(10):1818–1825.

Arima, H., Miyajib, T., Irie, T., Hirayama, F., Uekamaa, K. 1998. Enhancing effect of hydroxy-propyl-P-cyclodextrin on cutaneous penetration and activation of ethyl 4-biphenylyl acetate in hairless mouse skin. *Eur J Pharm Sci.* 6:53–59.

Autio, P., Karjalainen, J., Risteli, L., Risteli, J., Kiistala, U., Oikarinen, A. 1996. Effects of an inhaled steroid (budesonide) on skin collagen synthesis of asthma patients in vivo. *Am J Respir Crit Care Med.* 153(3):1172–1175.

Ayala-Bravo, H.A., Quintanar-Guerrero, D., Naik, A., Kalia, Y.N., Cornejo-Bravo, J.M., Ganem-Quintanar, A. 2003. Effects of sucrose oleate and sucrose laureate on in vivo human stratum corneum permeability. *Pharm Res.* 20:1267–1273.

Barry, B.W. 1983. Dermatological formulations: percutaneous absorption. In: Swarbick, J., ed. *Drugs and the Pharmaceutical Sciences.* Marcel Dekker, Inc., New York and Basel, p. 202.

Bashir, S.J., Chew, A.L., Anigbogu, A. 2001. Physical and physiological effects of stratum corneum tape stripping. *Skin Res Technol.* 7:40–48.

Bashir, S.J., Dreher, F., Chew, A.L., Zhai, H., Levina, C., Stern, R., Maibach, H.I. 2005. Cutaneous bioassay of salicylic acid as a keratolytic. *Int J Pharm.* 292:187–194.

Benfeldt, E., Hansen, S.H., Volund, A., Menne, T., Shah, V.P. 2007. Bioequivalence of topical formulations in humans: Evaluation by dermal microdialysis sampling and the dermatopharmacokinetic method. *J Invest Dermatol.* 127:170–178.

Benfeldt, E., Serup, J., Menné, T. 1999. Microdialysis vs suction blister technique for in vivo sampling of pharmacokinetics in the human dermis. *Acta Derm Venereol.* 79(5):338–342.

Bommannan, D., Potts, R.O., Guy, R.H. 1990. Examination of stratum corneum barrier function in vivo by infrared spectroscopy. *J Invest Dermatol.* 95:403–408.

Borg, N., Götharson, E., Benfeldt, E., Groth, L., Ståhle, L. 1999. Distribution to the skin of penciclovir after oral famciclovir administration in healthy volunteers: Comparison of the suction blister technique and cutaneous microdialysis. *Acta Derm Venereol.* 79(4):274–277.

Brian, K.R., Green, D.M., Lalko, J., Api, A.M. 2007. In-vitro human skin penetration of the fragrance material geranyl nitrile. *Toxicol In Vitro* 21:133–138.

Bucks, D.A., Hostynek, J.J., Hinz, R.S., Guy, R.H. 1993. Uptake of two zwitterionic surfactants into human skin in vivo. *Toxicol Appl Pharmacol.* 120:224–227.

Caron, D., Queille-Roussel, C., Shah, V.P., Schaefer, H. 1990. Correlation between the drug penetration and the blanching effect of topically applied hydrocortisone creams in human beings. *J Am Acad Dermatol.* 23:458–462.

Chao, Y.Ch., Nylander-French, L.A. 2004. Determination of keratin protein in a tape-stripped skin sample from jet fuel exposed skin. *Ann Occup Hyg.* 48(1):65–73.

Chi, S.Ch., Do, K., Tan, H.K., Chun, H.W. 1996. Anti-inflammatory and analgesic transdermal gel, *United States Patents*, Patent number 5,527,832.

Choi, M.J., Zhai, H., Kim, J-H., Maibach, H.I. 2008.Tape stripping method versus stratum corneum. In: Zhai, H., Wilhelm, K.P., Maibach, H.I. (eds.), *Dermatotoxicology*, 7th edn. CRC Press, Boca Raton, FL, pp. 327–337.

Coderch, L., de Pera, M., Perez-Cullell, N., Estelrich, J., de la Maza, A., Parra, J.L. 1999. The effect of liposomes on skin barrier structure. *Skin Pharmacol Appl Skin Physiol.* 12(5):235–246.

Conner, D.P.: Differences in DPK Methods. http://www.fda.gov/ohrms/dockets/ac/01/slides/3804s2_05_conner/index.htm, Advisory Committee for Pharmaceutical Sciences Meeting, Center for Drug Evaluation and Research (CDER), Food and Drug Administration (FDA), Rockville, MD, November 29, 2001.

Curdy, C., Kalia, Y.N., Naïk, A., Guy, R.H. 2001. Piroxicam delivery into human stratum corneum in vivo: Iontophoresis versus passive diffusion. *J Control Rel.* 76:73–79.

Darlenski, R., Sassning, S., Tsankov, N., Fluhr, J.W. 2009. Non-invasive in vivo methods for investigation of the skin barrier physical properties. *Eur J Pharm Biopharm.* 72(2):295–303.

De Jongh, C.M., Jakasa, I., Verberk, M.M., Kezic, S. 2006. Variation in barrier impairment and inflammation of human skin as determined by sodium lauryl sulphate penetration rate. *Br J Dermatol.* 154:651–657.

Dias, M., Naik, A., Guy, R.H., Hadgraft, J., Lane, M.E. 2008. In vivo infrared spectroscopy studies of alkanol effects on human skin. *Eur J Pharm Biopharm.* 69(3):1171–1175.

Ellias, P.M. 1991. Epidermal barrier function: Intercellular lamellar lipid structures, origin, composition and metabolism, *J Control Rel.* 15:199–208.

Escobar-Chávez, J.J., López-Cervantes, M., Naïk, A., Kalia, Y.N., Quintanar-Guerrero, D., Ganem-Quintanar, A. 2006. Applications of the thermoreversible Pluronic F-127 gels in pharmaceutical formulations, *J Pharm Pharmaceut Sci.* 9(3):339–358.

Escobar-Chávez, J.J., Melgoza-Contreras, L.M., López-Cervantes, M., Quintanar-Guerrero, D., Ganem-Quintanar, A. 2009. The tape stripping technique as a valuable tool for evaluating topical applied compounds. In: Caldwell, G.W., Atta-ur-Rahman, Yan, Z., Iqbal Choudhary, M. (eds.), *Frontiers in Drug Design & Discovery*. Bentham Science Publishers, Vol. 4. 189–227.

Escobar-Chávez, J.J., Merino-Sanjuán, V., López-Cervantes, M. et al. 2008. The tape-stripping technique as a method for drug quantification in skin. *J Pharm Pharmaceut Sci.* 11(1):104–130.

Escobar-Chávez, J.J., Quintanar-Guerrero, D., Ganem-Quintanar, A. 2005. *In vivo* skin permeation of sodium naproxen formulated in PF-127 gels: Effect of Azone® and Transcutol®, *Drug Develop Ind Pharm.* 31:447–454.

Esposito, E., Cortesi, R., Drechsler, M., Paccamiccio, L. et al. 2005. Cubosome dispersions as delivery systems for percutaneous administration of indomethacin. *Pharm Res.* 22(12):2163–2173.

Fang, J.Y., Leu, Y.L., Wang, Y.Y., Tsai, Y.H. 2002. *In vitro* topical application and *in vivo* phramacodynamic evaluation of nonivamide hydrogels using Wistar rat as an animal model, *Eur J Pharm Sci.* 15(5):417–423.

Fluhr, J.W., Gloor, M., Lehmann, L. 1999. Glycerol accelerates recovery of barrier function in vivo. *Acta Derm Venereol (Stockh).* 79:418–21.

Forslind, B.A. 1994. Domain mosaic model of skin barrier. *Acta Derm Venereol.* 74:1–6.

Franz, T.J. 1975. Percutaneous absorption on the relevance of in vitro data, *J Invest Dermatol.* 1975, 64:190–195.

Franz, T.J. Study #1, Avita Gel 0.025% vs Retin-A Gel 0.025%, Advisory committee for pharmaceutical sciences meeting, Center for Drug Evaluation and Research (CDER), Food and Drug Administration (FDA), Rockville, MD, November 29, 2001.

Fresno-Contreras, M.J., Jiménez-Soriano, M.M., Ramírez-Diéguez, A. 2005. In vitro percutaneous absorption of all-*trans* retinoic acid applied in free form or encapsulated in stratum corneum lipid liposomes. *Int J Pharm.* 297:134–145.

Ganem-Quintanar, A., Silva-Alvarez, M., Alvarez-Roman, R., Casas-Alancaster, N., Cazares-Delgadillo, J., Quintanar-Guerrero, D. 2006. Design and evaluation of a self—Adhesive naproxen-loaded film prepared from a nanoparticle dispersión. *J Nanosci Nanotechnol.* 6(9–10):3235–3241.

Ghadially, R., Brown, B.E., Sequeira-Martin, S.M. 1995.The aged epidermal permeability barrier. Structural, functional, and lipid biochemical abnormalities in humans and a senescent murine model. *J Clin Invest.* 95:2281–2290.

Goetz, N., Kaba, G., Good, D., Hussler, G., Bore, P. 1988. Detection and identification of volatile compounds evolved from human hair and scalp using headspace gas chromatography. *J Soc Cosmet Chem.* 39:1–13.

Groth, S., Staberg, B. 1984. Suction blister of the skin: A compartment with physiological, interstitium-like properties. *Scand J Clin Lab Invest.* 44:311–316.

Grundmann-Kollmann, M., Podda, M., Bräutigam, L., Hardt-Welnelt, K., Ludwing, R.J., Gelsslinger, G., Kaufmann, R., Tegeder, I. 2002. Spatial distribution of 8-methoxypsoralen penetration into human skin after systemic or topical administration. *Br J Clin Pharmacol.*54:535–539.

Guy, R.H., Hadgraft, J. 2003. *Transdermal Drug Delivery.* New York: Marcel Dekker, Inc., pp. 1–23.

Haaverstad, R., Romslo, I., Larsen, S., Myhre, H.O. 1996. Protein concentration of subcutaneous interstitial fluid in the human leg. A comparison between the wick technique and the blister suction technique. *Int J Microcirc Exp.* 16(3):111–117.

Hadgraft, J. 2001. Skin, the final frontier. *Int J Pharm.* 224 (1–2):1–18.

Hara, M., Verkman, A.S. 2003. Glycerol replacement corrects defective skin hydration, elasticity, and barrier function in aquaporin-3-deficient mice. *PNAS.* 100(12):7360–7365.

Herfst, M.J., van Rees, H. 1978. Suction blister fluid as a model for interstitial fluid in rats. *Arch Dermatol Res.* 263:325–324.

Herkenne, C., Alberti, I., Naik, A., Kalia, Y.N., Mathy, F.X., Préat, V., Guy, R.H. 2008. In vivo methods for the assessment of topical drug bioavailability. *Pharm Res.* 25(1):87–103.

Herkenne, C., Naik, A., Kalia, Y.N., Hadgraft, J., Guy, R.H. 2006. Pig ear skin ex vivo as a model for in vivo dermatopharmacokinetic studies in man. *Pharm Res*. 23(8):1850–1856.

Herkenne, C., Naik, A., Kalia, Y.N., Hadgraft, J., Guy, R.H. 2007. Dermatopharmacokinetic prediction of topical drug bioavailability in vivo. *J Invest Dermatol*. 127:887–894.

Higo, N., Naik, A., Bommannan, D.B., Potts, R.O., Guy, R. H. 1993.Validation of reflectance infrared spectroscopy as a quantitative method to measure percutaneous absorption in vivo. *Pharm Res*. 10:1500–1505.

Hostýnek, J.J., Dreher, F., Maibach, H.I. 2006. Human stratum corneum penetration by copper: In vivo study after occlusive and semi-occlusive application of the metal as powder. *Food Chem Toxicol*. 44:1539–1543.

Hostýnek, J.J., Dreher, F., Pelosi, A. 2001. Human stratum corneum penetration by nickel. In vivo study of depth distribution after occlusive application of the metal as powder. *Acta Derm Venereol (Stockh)*. 212:5–10.

Howes, D., Guy, R.H., Hadgraft, J., Heylings, J. et al. 1996. Methods for assessing percutaneous absorption. The report and recommendations of ECVAM workshop 13. *ATLA* 24:81–106.

Huang, Y.C., Lesko, L., Schwartz, P., Shah, V.P., Williams, R. 1993. Topical and transdermal generic drug products: Regulatory issues and resolution. In Brain, K.R., Walters, K.A., James, V.J. (eds.), *Prediction of Percutaneous Penetration*, Vol 3B. STS, Cardiff, pp. 463–472.

Ihlberg, L., Haukipuro, K., Risteli, L., Oikarinen, A., Kairaluoma, M.I., Risteli, J. 1993 Collagen synthesis in intact skin is suppressed during wound healing. *Ann Surg*. 217:397–403.

Inoue, J. and Aramaki, Y. 2007. Toll-like receptor-9 expression induced by tape-stripping triggers on effective immune response with CpG-oligodeoxynucleotides. *Vaccine* 25(6):1007–1013.

Inoue, J., Yotsumoto, S., Sakamoto, T., Tsuchiya, S., Aramaki, Y. 2005. Changes in immune responses to antigen applied to tape-stripped skin with CpG oligodeoxynucleotide in mice. *J Control Rel*. 108:294–305.

Jacobi, U., Tassopoulos, T., Surber, C., Lademann, J. 2006. Cutaneous distribution and localization of dyes affected by vehicles all with different lipophilicity. *Arch Dermatol Res*. 297:303–310.

Jakasa, I., Calkoen, F., Kezic, S. 2004. Determination of polyethylene glycols of different molecular weight in the stratum corneum. *J Chromatogr B*. 811:177–182.

Jakasa, I., Verbek, M.M., Esposito, M., Bos, J.D., Kezic, S. 2007. Altered penetration of polyethylene glycols into uninvolved skin of atopic dermatitis patients. *J Investigative Dermatol*. 127:129–134.

Jarvis, C.A., McGuigan, C., Heard, C.M. 2004. In vitro delivery of novel, highly potent anti-Varicella Zoster virus nucleoside analogues to their target site in the skin. *Pharm Res*. 21(6):914–919.

Kalia, Y.N., Alberti, I., Naïk, A., Guy, R.H. 2001. Assessment of topical bioavailability in vivo: The importance of stratum corneum thickness. *Skin Pharmacol Appl Skin Physiol*. 14(1):82–86.

Kaliyadan, F., Manoj, J., Venkitakrishnan, S. 2008. Using a microdermabrasion machine as a suction blister device. *Indian J Dermatol Venereol Leprol*. 74:392–393.

Kattan El, A.F., Asbill, C.S., Kim, N., Michniak, B.B. 2000. Effect of formulation variables on the percutaneous permeation of ketoprofen from gel formulations, *Drug Deliv*. 7(3):147–153.

Keiko, N., Toshiko Y. 1987. Volatilization of menthol, camphor and methyl salicylate from analgesic anti-inflammatory cataplasms and plasters. *Yakuzaigaku* 47:168–175.

Kiistala, R. 1992. Cholinergic and adrenergic sweating in atopic dermatitis. *Acta Derm Venereol*. 72(2):106–108.

Kiistala, U. 1968. Suction blister device for separating viable epidermis from dermis. *J Invest Dermatol*. 50:129–137.

Kiistala, R., Kiistala, U., Parkkinen, M.U., Mustakallio, K.K. 1984. Local sweat stimulation with the skin prick technique. *Acta Derm Venereol*. 64(5):384–388.

King, C.S., Barton, S.P., Nicholls, S., Marks, R. 1979. The change in properties of the stratum corneum as a function of depth. *Br J Dermatol*. 100:165–172.

Kondo, H., Ichikawa, Y., Imokawa, G. 1998. Percutaneous sensitization with allergens through barrier-disrupted skin elicits a Th2-dominant cytokine response. *Eur J Immunol*. 28:769–779.

Kreilgaard, M. 2002. Assessment of cutaneous drug delivery using microdialysis. *Adv. Drug Deliv Rev*. 54(1):S99–S121.

Królicki, A. 2002. Skin penetration of sulfamethoxazole and trimethoprim after oral adminis-tration. *Ann Acad Med Stetin*. 48:59–73.

Lademann, J., Jacobi, U., Surber, C., Weigmann, H.J., Fluhr, J.W. 2009. The tape stripping pro-cedure-evaluation of some critical parameters. *Eur J Pharm Biopharm*. 72(2):317–323.

Lademann, J., Otberg, N., Richter, H. 2001. Investigation of follicular penetration of topically applied substances. *Skin Pharmacol Appl Skin Physiol*. 14(1):17–22.

Lademann, J., Otberg, N., Richter, H., Jacobi, U., Schaefer, H., Blume-Peytavi, U., Sterry, W. 2003. Follicular penetration. An important pathway for topically applied substances. *Der Hautarzt*. 54:321–323.

Lademann, J., Schaefer, H., Otberg, N., Teichmann, A., Blume-Peytavi, U., Sterry, W. 2004a. Penetration of microparticles into human skin. *Der Hautarzt*. 55:1117–1119.

Lademann, J., Schanzer, S., Richter, H., Pelchrzim, R.V., Zastroe, L., Golz, K., Sterry, W. 2004b. Sunscreen application at the beach. *J Cosmet Dermatol*. 3(2):62–68.

Lboutounne, H., Chaulet, J.F., Ploton, C., Falson, F., Pirot, F. 2002. Sustained ex vivo skin antiseptic activity of chlorhexidine in poly(Є-caprolactone) nanocapsule encapsulated form and as a digluconate. *J Control Rel*. 82:319–334.

Leivo, T., Arjomaa, P., Oivula, J., Vesterinen, M., Kiistala, U., Autio, P., Oikarinen, A. 2000. Differential modulation of transforming growth factors-beta by betamethasone-17-valerate and isotretinoin: Corticosteroid decreases and isotretinoin increases the level of transforming growth factor-beta in suction blister fluid. *Skin Pharmacol Appl Skin Physiol*. 13(3–4):150–156.

Liaw, J., Lin, Y.-Ch. 2000. Evaluation of poly(ethylene oxide)-poly(propylene oxide)-poly(ethylene oxide) (PEO-PPO-PEO) gels as a release vehicle for percutaneous fen-tanyl, *J Control Rel*. 68:273–282.

Liu, L.-J., Watabe, S., Yang, J., Hamajima, K., Ishii, N., Hagiwara, E., Onari, K., Xin Q.-K., Okuda, K. 2001. Topical application of HIV DNA vaccine with cytokine-expression plasmids induces strong antigen-specific immune responses. *Vaccine*. 20(1–2):42–48.

Lodén, M., Akerstrom, U., Lindahl, K., Berne, B. 2004. Bioequivalence determination of topi-cal ketoprofen using a dermatopharmacokinetic approach and excised skin penetration. *Int J Pharm*. 284:23–30.

Löffler, H., Dreher, F., Maibach, H.I. 2004. Stratum corneum adhesive tape stripping: Influence of anatomical site, application pressure, duration and removal. *Br J Dermatol*. 151:746–752.

Löffler, H., Weimer, C., Dreher, F., Maibach, H.I. 2008. Parameters influencing stratum cor-neum removal by tape stripping. In: Zhai, H., Wilhelm, K.P., Maibach, H.I., (eds.), *Dermatotoxicology*, 7th edn. CRC Press, Boca Raton, FL, pp. 339–342.

Lotte, C., Wester, R.C., Rougier, A., Maibach, H.I. 1993. Racial differences in the *in vivo* percutaneous absorption of some organic compounds: A comparison between black, Caucasian and Asian subjects. *Arch Dermatol Res*. 284:456–459.

Lucker, G.P., van de Kerkhof, P.C., van Dijk, M.R., and Steijlen, P.M. 1994. Effect of topical calcipotriol on congenital ichthyoses. *Br J Dermatol*. 131(4):546–550.

Lücker, P.W., Beubler, E., Kukovetz, W.R., Ritter, W. 1984. Retention time and concentration in human skin of bifonazole and clotrimazole. *Dermatologica* 169:51–55.

Lundgren, L., Skare, L., Lidén, C. 2006. Measuring dust on skin with a small vacuuming sampler—A comparison with other sampling techniques. *Ann Occup Hyg.* 50(1):95–103.

Makki, S., Treffel, P., Humbert, P., Agache, P. 1991. High performance liquid chromatographic determination of citropten and bergapten in suction blister fluid after solar product application in humans. *J Chromatogr.* 563:407–413.

Marttin, E., Neelissen-Subnel, M.T.A, De Haan, F.H.N., Boddé, H.E. 1996. A critical comparison of methods to quantify stratum corneum removed by tape-stripping. *Skin Pharmacol.* 9:69–77.

Mattorano, D.A., Kupper, L.L., Nylander-French, L.A. 2004. Estimating dermal exposure to jet fuel (naphthalene) using adhesive tape strip samples. *Ann Occup Hyg.* 48(2):139–146.

Mavon, A., Miguel, C., Lejeune, O., Payre, B., Moretto, P. 2007. In vitro percutaneous absorption and in vivo stratum corneum distribution of an organic and mineral sunscreen. *Skin Pharmacol Physiol.* 20(1):10–20.

Mazzei, T., Tonelli, F., Novelli, A., Ficari, F., Mazzoni, C., Anastasi, A., Periti, P. 1994. Penetration of cefotetan into suction skin blister fluid and tissue homogenates in patients undergoing abdominal surgery. *Antimicrob Agents Chemother.* 38:2221–2223.

Miyazaki, S., Yokouchi, Ch., Nakamura, T., Hashiguchi, N., Hou, W.M., Takada, M. 1986. Pluronic F-127 gels as a novel vehicle for rectal administration of indomethacin, *Chem Pharm Bull.,* 34:1801–1808.

Morgan, C.J., Renwick, A.G., Friedmann, P.S. 2003. The role of stratum corneum and dermal microvascular perfusion in penetration and tissue levels of water-soluble drugs investigated by microdialysis. *Br J Dermatol.* 148:434–443.

Moser, K., Kriwet, K., Naik, A., Kalia, Y.N., Guy, R. H. 2001. Passive skin penetration enhancement and its quantification in vitro. *Eur J Pharm Biopharm.* 52:103–112.

Mouton, J.W., Horrevorts, A.M., Mulder, G.H., Prens, E.P., Michel, M.F. 1990. Pharmacokinetics of ceftazidime in serum and suction blister fluid during continuous and intermittent infusions in healthy volunteers. *Antimicrob Agents Chemother.* 34:2307–2311.

Müller, M., Brunner, M., Schmid, R., Putz, E.M., Schmiedberger, A., Wallner, I., Eichler, G. 1998. Comparison of three different experimental methods for the assessment of peripheral compartment pharmacokinetic in humans. *Life Sci.* 62:227–234.

Nangia, A., Camel, E., Berner, B. 1993. Influence of skin irritants on percutaneous absorption. *Pharm Res.*10:1756–1759.

Nevill, A.M. 1994. The need to scale for differences in body size and mass: And explanation of Klieber's 0.75 mass exponent. *Am Physiol Soc.* 2870–2873.

Nylander-French, L.A. 2000. A tape-stripping method for measuring dermal exposure to multifunctional acrylates. *Ann Occup Hyg.* 44:645–651.

Ohman, H., Vahlquist, A. 1994. In vivo studies concerning a pH gradient in human stratum corneum and upper epidermis. *Acta Derm Venereol (Stockh).* 74:375–379.

Oikarinen, A., Autio, P., Kiistala, U., Risteli, L., Risteli, J. 1992. A new method to measure type I and III collagen synthesis in human skin in vivo. Demonstration of decreased collagen synthesis after topical glucocorticoid treatment. *J Invest Dermatol.* 98:220–225.

Oikarinen, A., Savolainen, E.R., Tryggvason, K., Foidart, J.M., Kiistala, U. 1982. Basement membrane components and galactosylhydroxylysyl glucosyltransferase in suction blisters of human skin. *Br J Dermatol.* 106:257–266.

Olvera-Martínez, B.I., Cazares-Delgadillo, J., Calderilla-Fajardo, S.B., Villalobos-García, R., Ganem-Quintanar, A., Quintanar-Guerrero, D. 2005. Preparation of polymeric nanocapsules containing octyl methoxycinnamate by the emulsification–diffusion technique: Penetration across the stratum corneum. *J Pharm Sci.* 94:1552–1559.

Padula, C., Colombo, G., Nicoli, S., Catellani, P.L., Massimo, G., Santi, P. 2003. Bioadhesive film for the transdermal delivery of lidocaine: In vitro and in vivo behaviour. *J Control Rel.* 88:277–285.

Park, T.-H., Park, C.-H., Ha, S.-K., Lee, S.-H., Song, K.-S., Lee, H.-Y., Han, D.-S. 1995. Dry skin (xerosis) in patients undergoing maintenance haemodialysis: The role of decreased sweating of the eccrine sweat gland. *Nephrol Dial Transplant.*, 10(12):2269–2273.

Pechere, M., Krischer, J., Remondat, C. 1999. Malassezia spp. Carriage in patients with seborrheic dermatitis. *J Dermatol.* 26:558–561.

Pechere, M., Remondat, C., Bertrand, C. 1995. A simple quantitative culture of Malassezia spp. in HIV-positive persons. *Dermatology.* 191:348–349.

Pellanda, C., Ottiker, E., Strub, C., Figueiredo, V., Rufli, T., Imanidis, G., Surber, C. 2006. Topical bioavailability of triamcinolone acetonide: Effect of dose and application frequency. *Arch Dermatol Res.* 298:221–230.

Pershing, L.K. Bioequivalence assessment of three 0.025% tretinoin gel products: Dermatopharmacokinetic vs clinical trial methods, advisory committee for pharmaceutical sciences meeting, Center for Drug Evaluation and Research (CDER), Food and Drug Administration (FDA), Rockville, MD, November 29, 2001.

Pershing, L.K., Silver, B.S., Krueger, G.G., Shah, V.P., Skelley, J.P. 1992. Feasibility of measuring the bioavailability of topical betamethasone dipropionate in commercial formulations using drug content in skin and a skin blanching bioassay. *Pharm Res.* 9:45–51.

Pinkus, H. 1951. Examination of the epidermis by the strip method of removing horny layers. I. Observation on thickness of the horny layer, and on mitotic activity after stripping. *J. Invest. Dermatol.* 16:383–386.

Pinkus H. 1966. Tape stripping in dermatological research. A review with emphasis on epidermal biology. *G Ital Dermatol Minerva Dermatol.* 107(5):1115–26.

Pirot, F., Kalia, Y.N., Stinchcomb, A.L., Keating, G., Bunge, A., Guy, R.H. 1997. Characterization of the permeability barrier of human skin in vivo. *Proc Natl Acad Sci USA.* 94:1562–1567.

Potts, R.O., Francoeur, M.L. 1991. The influence of stratum corneum morphology on water permeability. *J Invest Dermatol.* 96:495–499.

Potts, R.O., Guy, R.H. 1991. Predicting skin permeability. *Pharm Res.* 9(5):663–669, (1992).

Pragst, F., Auwarter, V., Kiessling, B., Dyes, C. 2004. Wipe-test and patch-test for alcohol misuse based on the concentration ratio of fatty acid ethyl esters and squalene CFAEE/CSQ in skin surface lipids. *Forensic Sci Int.* 143:77–86.

Punnonen, K., Autio, P., Kiistala, U., Ahotupa, M. 1991. In vivo effects of solar-simulated ultraviolet iurradiation on antioxidant enzymes and lipid peroxidation in human epidermis. *Br J Dermatol.* 125(1):18–20.

Reddy, M.B., Stinchcomb, A.L., Guy, R.H., Bunge, A.L. 2002. Determining dermal absorption parameters in vivo from tape strip data. *Pharm Res.* 19:292–298.

Rosado, C., Rodrigues, L.M. 2003. Solvent effects in permeation assessed in vivo by skin surface biopsy. *BMC Dermatol.* 3:1–6.

Rougier, A., Dupuis, D., Lotte, C., Roguet, R., Wester, R.C., Maibach, H.I. 1986. Regional variation in percutaneous absorption in man: Measurement by the stripping method. *Arch Dermatol Res.* 278:465–469.

Sarveiya, V., Risk, S., Benson, H.A.E. 2004. Liquid chromatographic assay for common sunscreen agents: Application to in vivo assessment of skin penetration and systemic absorption in human volunteers. *J Chromatogr B.* 803:225–231.

Schafer, U. 2001. In: Bronaugh, R. L., Maibach, H. I. (eds.), *Topical Absorption of Dermatological Products.* Marcel Dekker, New York, Basel, p. 544.

Schrolnberger, C., Brunner, M., Mayer, B.X., Eichler, H.G., Müller, M. 2001. Application of the minimal trauma tissue biopsy to transdermal clinical pharmacokinetic studies. *J Control Release.* 75:297–306.

Shah, V.P. 1995. Challenges in evaluating bioequivalence of dermatological drug products. In: Surber, C. Elsner, P., J. Birhcer, A. (eds.), *Exogenous Dermatology. Current Problems in Dermatology.* Karger, Basel, Switzerland, pp. 152–157.

Shah, V.P. 1998. Topical Dermatological Drug Product NDAs and ANDAs-*In Vivo* Bioavailability, Bioequivalence, *In Vitro* Release and Associated Studies, US Department of Health and Human Services, Rockville, MD.

Shah, V.P., Flynn, G.L., Guy, R.H. et al. 1991. In vivo percutaneous penetration/absorption. *Pharm Res.*, 8:1071–1075.

Shah, V.P., Flynn, G.L., Yacobi, A., Maibach, H.I. et al. 1998. Bioequivalence of topical dermatological dosage forms-methods of evaluation of bioequivalence. *Pharm Res.* 15:167–171.

Shah, V.P., Hare, D., Dighe, S.V., Williams, R.L. 1993. Bioequivalence of topical dermatological products. In: Shah, V.P., Maibach, H.I. (eds.), *Topical Drug Bioavailability, Bioequivalence and Penetration.* Plenum, New York, pp. 393–413.

Sheth, N.V., McKeough, M.B., Spruance, S.L. 1987. Measurement of the stratum corneum drug reservoir to predict the therapeutic efficacy of topical iododeoxyuridine for herpes simplex virus infection. *J Invest Dermatol.* 89:598–602.

Shin, S.C., Cho, C.W., Oh, I.J. 2001. Effects of non ionic surfactants as permeation enhancers towards piroxicam from the poloxamer gel through rat skins, *Int J Pharm.* 222(2):199–203.

Strid, J., Hourihane, J., Kimber, I., Callard, R., Strobel, S. 2004. Disruption of the stratum corneum allows potent epicutaneous immunization with protein antigens resulting in a dominant systemic Th2 response. *Eur J Immunol.* 34:2100–2109.

Surakka, J., Johnsson, S., Rosen, G. 1999. A method for measuring dermal exposure to multi-functional acrylates. *J Environ Monit.* 1:533–540.

Surakka, J., Lindh, T., Rosen, G. 2000. Workers' dermal exposure to UV-curable acrylates in the furniture and parquet industry. *Ann Occup Hyg.* 44:635–644.

Surber, C., Schwarb, F.P., Fmith, E.W. 1999. Tape stripping technique. In: Bronough, H., Maibach, H.I. (eds.), *Percutaneous Absorption—Drug—Cosmetics—Mechanisms—Methodology*, 3rd edn. Marcel Dekker, New York, pp. 395–409.

Surber, C., Wilhelm, K.P., Bermann, D., Maibach, H.I. 1993. In vivo skin penetration of acitretin in volunteers using three sampling techniques. *Pharm Res.* 10:1291–1294.

Svedman, P. 1996. Suction Blister Sampling. WO/1996/033768 Patent.

Svedman, P., Svedman, C. 1998. Skin mini-erosion sampling technique: Feasibility study with regard to serial glucose measurement. *Pharm Res.* 15:883–888.

Sylvestre, J.P., Díaz-Marín, C., Delgado-Charro, M.B., Guy, R.H. 2008. Iontophoresis of dexamethasonephosphate: Competition with chloride ions. *J Control Release.* 131(1):41–46.

Tegeder, I., Brautigam, L., Podda, M., Meier, S., Kaufmann, R., Geisslinger, G., Grundmann-Kollmann, M. 2002. Time course of 8-methoxypsoralen concentrations in skin and plasma after topical (bath and cream) and oral administration of 8-methoxypsoralen. *Clin Pharmacol Ther.* 71(3):153–161.

Teichmann, A., Jacobi, U., Ossadnik, M., Richter, H., Koch, S., Sterry, W., Lademann, J. 2005. Differential stripping: Determination of the amount of topically applied substances penetrated into hair follicles. *J Invest Dermatol.* 125:264–269.

Teichmann, A., Jacobi, U., Waibler, E., Sterry, W., Lademann, J. 2006. An in vivo model to evaluate the efficacy of barrier creams on the level of skin penetration of chemicals. *Contact Dermatitis.* 54:5–13.

Tettey-Amlalo, R.N.O., Kanfer, I., Skinner, M.F., Benfeldt, E., Verbeeck, R.K. 2009. Application of dermal microdialysis for the evaluation of bioequivalence of a ketoprofen topical gel. *Eur J Pharm Sci.* 36:219–225.

Tojo, K., Lee, A.C. 1989. A method for prediction of steady-state of skin penetration in vivo. *J Invest Dermatol.* 92:105–108.

Toll, R., Jacobi, U., Richter, H., Lademann, J., Schaefer, H., Blume-Peytavi, U. 2004. Penetration profile of microspheres in follicular targeting of terminal hair follicles. *J Invest Dermatol.* 123:168–176.

Touitou, E., Meidan, V.M., Horwtz, E. 1998. Methods for quantitative determination of drug localized in the skin. *J Control Release.* 56:7–21.

Toyoda, M., Nakamura, M., Nakagawa, H. 2007. Distribution to the skin of epinastine hydrochloride in atopic dermatitis patients. *Eur J Dermatol*. 17.

Treffel, P., Makki, S., Faivre, B., Humbert, P., Blanc, D., Agache, P. 1991. Citropten and bergapten suction blister fluid concentrations after solar product application in man. *Skin Pharmacol*. 4:100–108.

Tsai, H.C., Lin, C.Y., Sheu, H.M., Lo, Y.L., Huang, Y.H. 2003. Noninvasive characterization of regional variation in drug transport into human stratum corneum in vivo. *Pharm Res*. 20:632–638.

Tsai, J.-C., Shen, L.-C., Sheu, H.M., Lu, C.-C. 2003. Tape stripping and sodium dodecyl sulfate treatment increase the molecular weight cut-off of polyethylene glycol penetration across murine skin. *Arch Dermatol Res*. 295:169–174.

Tsai, J.-C., Weiner, N.D., Flynn, G.L., Ferry, J. 1991. Properties of adhesive tapes used for stratum corneum stripping, *Int J Pharm*. 72:227–231.

Vaillant, L., Autret, E., Marchand, S., Lorette, G., Grenier, B. 1991. Suction blisters technique in amikacin diffusion through interstitial fluid in cystic fibrosis. *Dev Pharmacol Ther*. 16:29–32.

Van der Merwe, D., Brooks, J.D., Gehring, R., Baynes, R.E., Monteiro-Riviere, N.A., Riviere, J.E. 2006. A physiologically based pharmacokinetic model of organophosphate dermal absorption. *Toxicol Sci*. 89(1):188–204.

Van der Molen, R.G., Spies, F., Van 't Noordende, J.M., Boelsma, E., Mommaas, A.M., Koerten, H.K. 1997. Tape stripping of human stratum corneum yields cell layers that originate from various depths because of furrows in the skin. *Arch Dermatol Res*. 289:514–518.

Van der Valk, P.G., Maibach, H.I. 1990. A functional study of the skin barrier to evaporative water loss by means of repeated cellophane-tape stripping. *Clin Exp Dermatol*. 15:180–182.

Van Voorst Vader, P.C., Lier, J.G., Woest, T.E. 1991. Patch tests with house dust mite antigens in atopic dermatitis patients: Methodological problems. *Acta Derm Venereol (Stockh)*. 71:301–305.

Vermeer, B.J., Reman, F.C., van Gent, C.M. 1979. The determination of lipids and proteins in suction blister fluid. *J Invest Dermatol*.73:303–305.

Vessby, B., Gustafson, S., Chapman, M.J., Hellsing, K., Lithell, H. 1987. Lipoprotein composition of human suction-blister interstitial fluid. *J Lipid Res*. 28:629–641.

Volden, G., Thorsrud, A.K., Bjornson, I., Jellum, E. 1980. Biochemical composition of suction blister fluid determined by high resolution multicomponent analysis (capillary gas chromatography—Mass spectrometry and two-dimensional electrophoresis). *J Invest Dermatol*. 75:421–424.

Wagner, H., Kostka, K.-H., Lehr, C.-M., Schaefer, U.F. 2002. Correlation between stratum corneum/water-partition coefficient and amounts of flufenamic acid penetrated into the stratum corneum. *J Pharm Sci*. 91(8):1915–1921.

Wagner, S. and Merfort, I. 2007. Skin penetration behaviour of sesquiterpene lactones from different *Arnica* preparations using a validated GC-MSD method. *J Pharm Biomed Anal*. 43(1):32–38.

Walker, J.S., Knihinicki, R.D., Seideman, P., Day, R.O. 1993. Pharmacokinetics of ibuprofen enantiomers in plasma and suction blister fluid in healthy volunteers. *J Pharm Sci*. 82(8):787–790.

Walters, K.A., Roberts, M.S. 2002. *Dermatological and Transdermal Formulations*. Marcel Dekker, Inc., New York, pp. 1–39.

Wang, Y.Y., Hong, C.T., Chiu, W.T., Fang, J.Y. 2001. *In vitro* and *in vivo* evaluations of topically applied capsaicin and nonivamide from hydrogels, *Int J Pharm*. 224(1–2):89–104.

Wiedersberg, S., Naik, A., Leopold, C.S., Guy, R.H. 2009. Pharmacodynamics and dermatopharmacokinetics of betamethasone 17-valerate: Assessment of topical bioavailability. *Br J Dermatol*. 160(3):676–686.

Wilhelm, K.P., Surber, C., Maibach, H.I. 1991. Effect of sodium lauryl sulfate- induced skin irritation on *in vivo* percutaneous penetration of four drugs. *J Invest Dermatol*. 97:927–32.

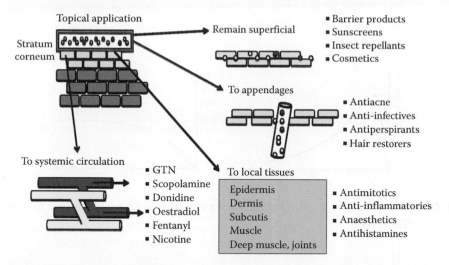

FIGURE 1.1 Schematic diagram showing various processes when the skin is exposed to different molecules. (Modified from Roberts, M.S. et al., Skin transport, In: Walters, K.A., editor, *Dermatological and Transdermal Formulations*, Marcel Dekker, New York, 2002, pp. 89–195.)

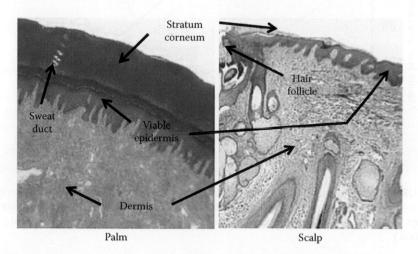

FIGURE 1.2 Structure of mammalian skin. (With permission of Allan F. Wiechmann, reproduction of slides 44 and supplemental slide 101 at www.ouhsc.edu/histology/, accessed March 2010.)

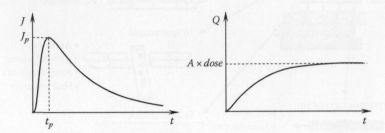

FIGURE 1.6 The flux of solute and the amount of a solute absorbed versus time for finite vehicle application.

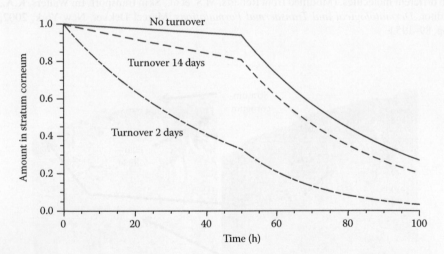

FIGURE 1.11 Amount remaining in SC reservoir of corticosterone with no desquamation (—), a normal epidermal turnover of 14 days (- - -) and a psoriatic epidermal turnover of 2 days (- – -.). (Modified from Roberts, M.S. et al., *Skin Pharmacol. Appl. Skin Physiol.*, 17, 3, 2004.)

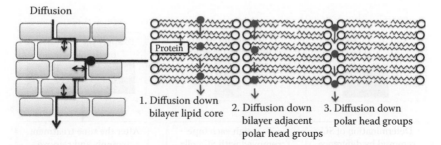

Diffusion

Protein

1. Diffusion down
bilayer lipid core

2. Diffusion down
bilayer adjacent
polar head groups

3. Diffusion down
polar head groups

FIGURE 1.12 Partition and diffusive processes in solute transport through the SC, assessing an intercellular lipid pathway. (Modified from Roberts, M.S. et al., Skin transport, In: Walters, K.A., editor, *Dermatological and Transdermal Formulations*, Marcel Dekker, New York, 2002, pp. 89–195.)

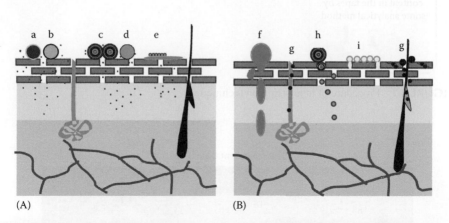

(A)   (B)

FIGURE 3.3 (A) *Potential mechanisms of prolonged drug delivery using novel drug delivery systems*: The microparticles ((a) reservoir and (b) matrix type), liposomes ((c) MLV and (d) LUV) adhere onto the skin surface and slowly release the drug. (e) Lipid nanoparticles or liposomes form a barrier layer and decrease the rate of drug delivery into skin. (B) *Potential mechanisms of enhanced drug delivery or drug targeting using novel drug delivery systems*: (f) The ultradeformable liposomes penetrate the stratum corneum and enter deeper layers due to the gradient in the water content. (g) Enhanced permeation or targeting of drugs to skin appendages and furrows in the stratum corneum. (h) MLVs lose peripheral layers during penetration across the stratum corneum. (i) The vesicular drug delivery systems permeabilize the lipid domain by fusion or by release of enhancer, thus enhancing the drug penetration.

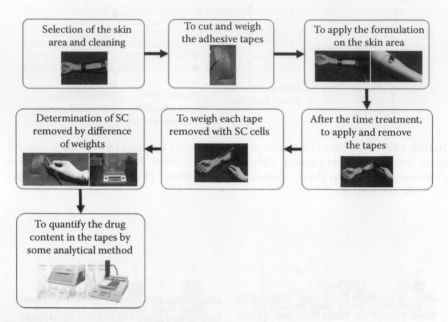

FIGURE 4.1 Procedure of tape-stripping technique.

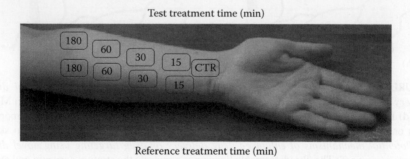

FIGURE 4.3 Assessment of topical BE of test and reference formulations. The SC is tape stripped immediately after each treatment time to determine the drug level in the barrier (CTR, control).

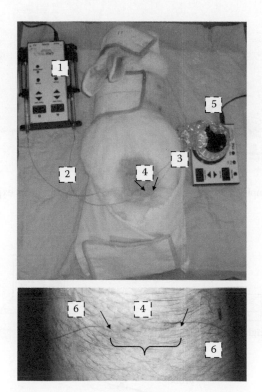

FIGURE 5.1 Typical microdialysis experimental setup in a rabbit model and in a human arm: (1) pump; (2) and (3) connecting tubes; (4) probe inside the skin; arrows indicate the inlet and outlet, respectively; (5) dialysate automatic fraction collector; and (6) tape to hold in place the probe.

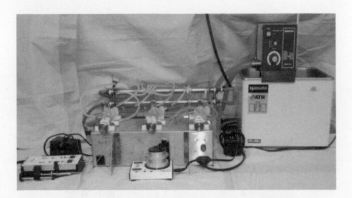

FIGURE 5.4 Example of an apparatus used to perform microdialysis experiments in vitro.

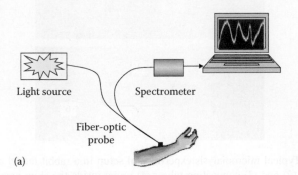

(a)

FIGURE 7.1 (a) A diffuse reflectance spectroscopy (DRS) system comprises a light source, a probe, and a computer-controlled spectrometer. The probe can be a bifurcated optical fiber, as depicted here, or an integrating sphere. The geometry of the probe is one of the parameters determining the detection depth in the tissue.

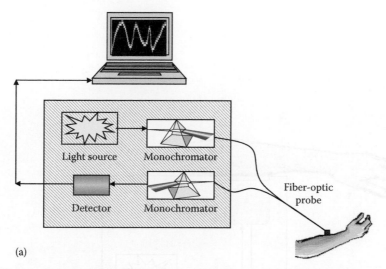

(a)

FIGURE 7.2 (a) A fluorescence spectroscopy (FS) system comprises a light source, a probe, and a sensitive detector. Double monochromators after the light source and just before the detector allow for the selection of excitation and emission wavelengths, respectively. The probe is typically a bifurcated optical fiber.

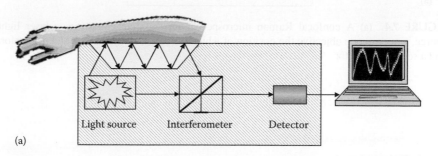

(a)

FIGURE 7.3 (a) An ATR-FTIR spectroscopy system comprises a laser light source, an internal reflectance element (typically ZnSe crystal) probe, and an interferometer with a liquid-nitrogen-cooled detector.

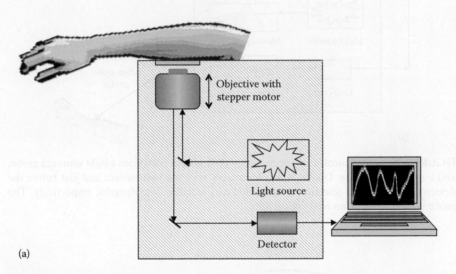

(a)

FIGURE 7.4 (a) A confocal Raman microspectroscopy system comprises a laser light source, a microscope objective the position of which is controlled by a z-axis stepper motor, and a sensitive detector.

5 Cutaneous Microdialysis

Grazia Stagni

CONTENTS

5.1 HISTORY

Microdialysis is commonly referred to as a sampling technique that uses tiny (micro) hollow probes (200–500 μm in diameter) equipped with a porous tubular segment (the dialysis membrane) that is permeable to compounds of size smaller than the pores by passive diffusion. Microdialysis was developed in order to sample in vivo endogenous compounds from the body with minimal tissue damage and under physiological conditions. The first use of the technique dates back to the 1960s [1] when Bito et al. used small dialysis sacs to sample amino acids in the brains of dogs. However, the technique as we use it today was mostly developed in the 1980s and 1990s and is still actively investigated for novel improvements and variations. This chapter focuses on the technique as applied to improve our understanding of drug dermatokinetics. Microdialysis in brain and other organs requires an invasive procedure for the insertion of the probe, whereas it is easily performed in the skin without the need of any surgical procedure. Clearly, this is a significant advantage.

5.2　BASICS OF CUTANEOUS MICRODIALYSIS

5.2.1　EQUIPMENT

Typically, a cutaneous microdialysis experiment requires equipment and an experimental setup similar to that shown in Figure 5.1 for rabbit and human models. The perfusate solution is stored in an appropriately sized syringe made of glass or disposable plastic that is inserted in a precision pump capable of delivering the fluid (the perfusate) at a very slow and accurately monitored flow rate. The syringe is connected to the probe via a tiny tube made of a inert material like Teflon. The tubing is then joined to the probe via a connector or glued to it. The probe could have different shapes [2]. However, for application in skin, the linear probe and the Y-shaped probe (Figure 5.2) are most commonly used. In both cases, the probe consists of three parts: the dialysis membrane, and the inlet and the outlet tubes made of an impermeable material that provides a rigid structure to the probe to facilitate

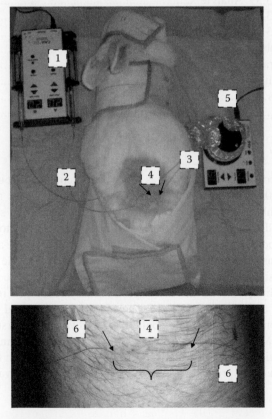

FIGURE 5.1　(See color insert.) Typical microdialysis experimental setup in a rabbit model and in a human arm: (1) pump; (2) and (3) connecting tubes; (4) probe inside the skin; arrows indicate the inlet and outlet, respectively; (5) dialysate automatic fraction collector; and (6) tape to hold in place the probe.

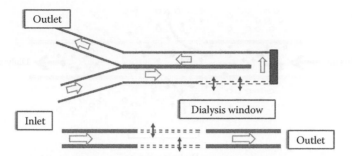

FIGURE 5.2 Most popular types of probe geometries used for cutaneous microdialysis. Top: Y-shaped probe. Bottom: linear probe.

the insertion process. The dialysis window is where the actual sampling process takes place. The length of the dialysis window, the material and thickness of the membrane, and the size of the pores are all factors that affect the performance of the technique and will be discussed separately. Finally, the outlet of the probe can either be placed directly in the collecting vial or can be connected via an inert tube to a fractions' collector. Possibly, the vial can fit in an HPLC autosampler for subsequent analysis. Sometimes, the outlet can be connected directly with the injector of an HPLC system [3]. It is important to ensure that the diameters of the tubing are consistent among the several segments or that they are of increasing size in order to avoid back pressure problems that may adversely affect the overall performance of the technique. For the same reason, it is advisable to place the collecting tube lower than the probe. After flowing through the dialysis window, the "perfusate" becomes the "dialysate." Usually the probe is inserted into the dermis of the subject via a sterile technique. Typically, a rigid guide, either a needle or a cannula, is inserted initially in the subject's skin, the probe is passed through it, and then the rigid guide is removed. Then the probe is connected to the pump and the sampler collector as described above. The inserted probe is secured to the body with a surgical tape. Every component of this experimental setup may affect the final performance of the technique and has to be carefully considered in the optimization of the experimental conditions.

5.2.2 THEORETICAL BACKGROUND AND THE CONCEPT OF RECOVERY

Microdialysis sampling occurs through the dialysis window. Molecules present in the dermis extracellular fluid (ECF) migrate across the pores of the dialysis membrane into the perfusing fluid by passive diffusion according to the concentration gradient. By the same principle, molecules present in the perfusate will move to the interstitial space (Figure 5.3). This process is mathematically modeled by Fick's first law [4]:

$$J = -D\frac{dC}{dx} \tag{5.1}$$

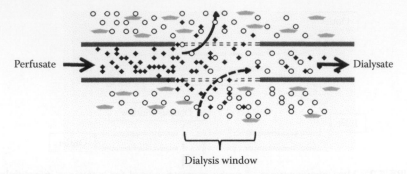

Dialysis window

FIGURE 5.3   The principle of microdialysis.

where

the flux (J) is the number of molecules diffusing across a unit cross section in a
   unit of time

dC/dx is the incremental change in concentration with distance (concentration
   gradient)

D is the diffusion coefficient

D depends on a variety of factors such as temperature, dialysis membrane characteristics, size and density of the membrane pores, and the diffusant's chemical nature.

Other more sophisticated mathematical models have been proposed to account
also for convection or lateral diffusion [5,6]. However, these phenomena are usually
considered negligible for the experimental conditions typically used in cutaneous
microdialysis.

In conditions of no flow, the concentration of each compound inside and outside the probe would be the same after allowing enough time to reach equilibrium.
However, in the most popular applications for skin sampling, the probe is continuously perfused, and diffusional equilibrium is never reached. Therefore, the concentration of the analyte is generally lower in the collected dialysate than that in the
interstitial fluid. The ratio between the concentration in the dialysate and that in the
periprobe fluid is called relative recovery [7]:

$$Recovery = \frac{[C]_{dialysate}}{[C]_{periprobe}} \tag{5.2}$$

Because of the conditions of nonequilibrium, 100% relative recovery is never
reached. Therefore, it is necessary to know the relative recovery in order to
calculate the actual periprobe concentration in in vivo experiments, by solving
Equation 5.2 for $[C]_{periprobe}$. Relative recovery can be estimated in vitro to obtain
a preliminary assessment of the feasibility of the microdialysis technique for the
molecules studied. However, in vitro recovery cannot be used to correct for recovery in vivo because the diffusivity characteristics of the ECF differ from those
of the solution used in vitro [8]. Nevertheless, in vitro experiments are useful for
a preliminary optimization of the technique since all the parameters affecting

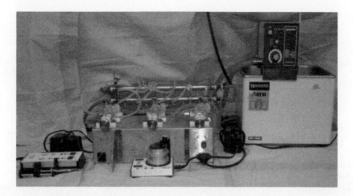

FIGURE 5.4 **(See color insert.)** Example of an apparatus used to perform microdialysis experiments in vitro.

recovery are the same, with the only exception of the diffusivity in the ECF. A typical experimental setup for in vitro microdialysis is shown in Figure 5.4. Here, the microdialysis probe is inserted in the inner chamber of a double-jacket glass cell in such a way that the membrane window lies exactly at the center of the chamber. The temperature of the circulating water in the cell jacket is maintained at 35°C to mimic the skin temperature and the solution in the cell is stirred continuously. The remaining equipment and setup are the same as those used in vivo (Figure 5.1).

5.2.3 FACTORS AFFECTING RECOVERY

The relative recovery is affected by several factors: perfusate composition and flow rate, membrane type and length, probe geometry, and the body tissue where the probe is implanted (Figure 5.5).

5.2.3.1 Perfusate

5.2.3.1.1 Composition
The fluid perfused throughout the microdialysis apparatus is called perfusate. It is usually a water-based solution with the only requirement that it be isotonic with the ECF to prevent loss or gain of water during the experiment that would affect the quantification process. This can be experimentally verified by measuring the volume of fluid collected [9]. For example, if the flow rate is 2 µL/min, the volume collected over a time period of 10 min is 20 µL. An experimental volume less than 20 µL indicates loss of water from the perfusate; a larger volume indicates gain of water from the ECF. In both cases, the periprobe environment is altered. Isotonicity protects against such a net exchange of water, whereas substances present in the perfusate will still move across the membrane according to the respective concentration gradients. In theory, the closer the composition of the perfusate is to that of the ECF the better in order to avoid alteration of the ECF ionic composition. This is particularly true for physiological studies. However, while some

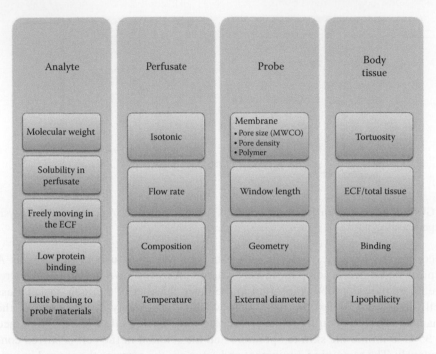

FIGURE 5.5   Factors affecting microdialysis recovery at a glance. (ECF = extracellular fluid.)

authors observed effects on the physiology of the brain [10] others reported no detectable evidence of alteration [11]. The solution of choice used for cutaneous microdialysis is Lactated Ringer's (LR) because it has a composition similar to that of the dermis ECF. If LR interferes with the analytical assay, then normal saline (0.9% NaCl) can be used [12]. In some cases, an isotonic phosphate buffer of pH 7.4 was used [13] to buffer the pH at physiological values. Sometimes, addition of a small amount (0.2%–0.5%) of albumin (BSA) is necessary to prevent the binding of the analytes to the apparatus material. These simple isotonic solutions are ideal to sample water soluble compounds that are freely moving in the extracellular space but are usually unsuitable to collect lipophilic or highly protein-bound substances [7,14]. In order to improve sampling of lipophilic or highly protein-bound substances, special perfusates have been used. For example, the addition of 4% BSA improved the recovery of oleic acid [15]. Muller et al. [16] reported that 7% BSA improved the recovery of nicotine and estradiol from the skin. They also reported that 7% was the highest practically viable concentration of BSA since greater amounts of albumin increased the viscosity of the perfusion fluid so much that it could not flow through the 0.12 mm internal diameter probe. The use of isotonic microemulsion gave results similar to those with 4% BSA for oleic acid [15]. The addition of binding agents such as cyclodextrin improved the recovery of hydrophobic tricyclic drugs [17]. Ward et al. [18] found that the use of Encapsin, a cyclodextrin, to improve the recovery of a highly lipophilic substance is comparable to perfusion with a microemulsion.

These methods improve significantly the recovery but the sample needs to be processed before injection into an HPLC, thus losing one of the most appealing advantages of microdialysis.

Other factors to be considered in the selection of the perfusate are the analytical specificity of the assay and possible chemical interactions with the analyte, such as complexation, precipitation, and so on. Moreover, the perfusate must be free of air bubbles that may decrease the recovery and increase the variability. Removal of air bubbles can be achieved by vacuum filtration.

5.2.3.1.2   Flow Rate

The rate at which the perfusate passes through the probe is a critical determinant of recovery. A slow flow rate allows more time for the molecules to cross into the perfusate and the relative recovery (Equation 5.2) is higher. Conversely, with a fast flow rate, the concentration of the analyte in the dialysate is lower although the total amount collected is greater [19]. Indeed, recovery can be expressed also as mass units removed per time interval. This is called absolute (or mass) recovery [19]. Figure 5.6 shows the relationships between relative recovery, absolute recovery, and flow rate. Increasing the flow rate increases the pressure gradient between the perfusate and the periprobe solution, thus forcing some solvent to migrate from inside the probe to outside. Consequently, the volume of dialysate collected is smaller than that calculated theoretically. Also, the pressure gradient hinders the passage of analyte molecules into the perfusate and lowers the recovery.

Usually, flow rates up to 5 μL/min are considered safe [20]. Within this boundary, the selection of the flow rate is determined mostly by practical factors such as the capability of the analytical equipment to accurately handle and analyze very small volumes of samples (<10 μL), and the sensitivity of the analytical assay.

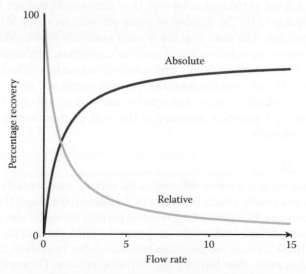

FIGURE 5.6   Absolute and relative recovery as a function of flow rate: absolute recovery increases with flow rate and relative recovery decreases with flow rate.

The availability of a pump that can deliver the perfusate accurately at a very low flow rate (0.1–0.5 μL/min) is also important. Optimization of the flow rate can be done in vitro, keeping in mind that usually in vivo recovery is less than the in vitro recovery.

5.2.3.1.3 Sample Collection Time

Directly related to the flow rate is the length of the sampling interval. A long sampling time may allow for the collection of a sufficient volume of sample to be handled with an ordinary equipment and for a better analytical quantification. It would therefore compensate for a very slow flow rate. In this case, it is important to evaluate the temporal resolution of the sampling collection with respect to the purpose of the study and the half-life of the analyte in the skin. For example, a 30 min sampling collection would not be appropriate for a drug with a half-life of less than 1 h.

5.2.3.2 Probe Membrane

5.2.3.2.1 Material

The tubular hollow dialysis fibers used for the probes are derived from those utilized in artificial kidneys. For an extensive review of dialysis membrane's chemistry and properties see Refs. [21,22]. In short, kidney fibers vary in inner diameter (180–220 μm), in the thickness of the walls (15–50 μm), and in the size and distribution of the pores. For microdialysis, thinner walls are preferred because they are more flexible and provide increased speed of the compounds' transport and decreased binding to the walls of the membrane during transport. Moreover, the pores are not perfectly straight channels that cross the membrane perpendicularly from side to side and may include some tortuosity. Also, the pore radius is not completely uniform and actually exhibits a Gaussian distribution. The narrower the distribution, the more sharply defined is the molecular weight (MW) cutoff of the membrane. The MW cutoff is defined as the molecular weight of compounds that are 90% rejected from the membrane [23]. The number of pores per unit area also affects the efficiency of the process. The most commonly used materials in microdialysis probes for skin applications are cellulose based (such as cuprophan, regenerated cellulose, or cellulose acetate), which are highly hydrophilic in nature or synthetic membranes (such as polyamide, polyarylethersulphone, polyethersulphone, polycarbonate, and polyacrylonitrile), which are more hydrophobic and can be affected by absorption of proteins. Table 5.1 reports a summary of the most popular membranes used in cutaneous microdialysis.

5.2.3.2.2 Length

The length of the dialysis window determines the surface area available for diffusion of the analytes and greatly affects the recovery. Intuitively, the longer the window the higher is the recovery. However, a long window permits the molecules just collected in the dialysate to diffuse back out to the periprobe fluid, particularly at very slow flow rates [9]. In addition, as the perfusate becomes richer in analyte, the concentration gradient decreases, thus limiting the diffusion process. Generally, membrane lengths of 1–10 mm are used in cutaneous microdialysis; however, up to 30 mm [24] membranes have been used in subcutaneous tissues. It is important that the dialysis

TABLE 5.1

Synopsis of the Characteristics of the Most Popular Membranes Used in Cutaneous Microdialysis

Material	Wall Thickness (µm)	ID[a] (µm)	MWCO[b]
Cuprophan (cellulose)	11	200	6,000
Regenerated cellulose	8	200	13,000 or 18,000
Cellulose acetate	9	232	5,000
Polyarylethersulfone (PAES)	50	500	20,000
Polyether sulfone (PES)	30 or 35	500	55,000 or 100,000
Polycarbonate	50	400	20,000
Polyacrylonitrile	50	240	29,000 or 50,000

Sources: Stenken, J.A. et al., *Anal. Chem.*, 65, 2324, 1993; Sun, L. and Stenken, J.A., *J. Pharm. Biomed. Anal.*, 33, 1059, 2003; Stenken, J.A. et al., *Anal. Chim. Acta*, 436, 21, 2001.

[a] Internal diameter.
[b] Molecular weight cutoff.

window is inserted in a completely homogeneous region of the tissue. Longer fibers may be difficult to insert precisely in the dermis for their entire length.

5.2.3.2.3   Geometry

The selection of the probe geometry is determined primarily by the location and type of tissue that is being investigated as well as by the size of the subject (e.g., rodents vs. humans). Usually, Y-shaped probes are employed to reach into deep tissues (such as the brain), whereas linear probes are more suitable for peripheral tissues such as skin, subcutaneous fat, or muscles. Probes can be custom made by the investigator or purchased. The commercial availability of a variety of differently shaped probes is on the rise and the number of companies manufacturing them is steadily increasing. The geometry of the probe may affect the speed at which the perfusate passes through the membrane, and, consequently, the recovery [7]. Some probe designs include a reinforcing stainless-steel wire support to strengthen the mechanical properties of the probe. Although the presence of the wire could theoretically alter the flow though the probe, it has been demonstrated [25] that the presence of the wire does not affect the experimental recovery of linear probes at the flow rates used for skin microdialysis.

5.2.3.3   Analyte

The most critical factor for the success of a microdialysis study is the suitability of the analyte for the technique. Generally, compounds with a molecular weight less than 1000, water soluble at physiological pH, and with no (or weak) bounding to ECF proteins and structures have the largest probability to be successfully collected by microdialysis with the types of membranes previously discussed. Microdialysis of larger molecules (like biologics) requires membranes with greater pore size.

Conversely, compounds that are not soluble in the perfusing fluid or are not freely moving in the ECF will not be collected. It is important to stress that a satisfactory recovery in vitro is not a sufficient predictor of in vivo recovery. There are a few cases (e.g., pentamidine [26] and triamcinolone acetonide [27]) of analytes that were successfully recovered in vitro but not recovered in skin after either transdermal or systemic administration.

5.2.3.4 Body Tissue

The characteristic of the body tissue in which the probe is implanted influences the recovery. The diffusion coefficient in Equation 5.2 is primarily determined by the tissue composition. While the dialysis membrane could theoretically hinder diffusion, this effect has been found to be experimentally negligible [5]. Consequently, analytes have different recoveries in different body tissues. Indeed, analytes may have different bounding affinity to the components of the interstitial fluid. In addition, the tortuosity of the tissue and the percentage of extracellular space compared to the total tissue affect the diffusivity of molecules, and, therefore, the recovery. In simple words, tortuosity is a measure of how difficult it is for a molecule to move within the ECF and to reach the dialysis membrane. Dermis and subcutaneous tissues are generally less tortuous than the brain. Therefore, even though successful microdialysis recovery in a tissue may be a good predictor of success in other tissues, recovery must always be measured experimentally in the tissue of interest.

5.2.3.5 Effect of Electric Current

Electric current is used to enhance transdermal delivery, such as in iontophoresis, and it stimulates physiological responses in the skin [28,29] (i.e., edema [30], erythema [31], and changes in pH [32]) as well as compensatory physiological feedback mechanisms [33] that may alter the environment surrounding the microdialysis probe and change the recovery or delivery properties of the probe. The effect of electric current on microdialysis recovery was studied in the human forearm by Stagni et al. [34] to assess whether these effects may cause a measurable change in the retrodialysis of a model compound (sodium fluorescein). Current was applied for two periods of 30 min each, separated by 30 min of no current. Changes in blood flow were monitored by laser Doppler flowmetry. Skin blood flow in response to iontophoresis was, on average, 570% higher than the control site. However, there was no difference in the recovery of fluorescein between the current active versus the control site. Holovics et al. [35] performed a similar study in hairless rats for a longer period of time (20 h). Also, in this study, no differences were observed between the microdialysis recoveries of lidocaine during iontophoresis compared to that without electric current. In conclusion, iontophoretic current does not influence intradermal microdialysis recovery.

5.3 IN VIVO PROBE CALIBRATION

Several methods for quantitative microdialysis in vivo have been developed and investigated in the past 20 years. In particular, three methods have gained wide spread acceptance: (1) the no-flux method, (2) the retrodialysis method, and (3) the internal standard method.

The no-flux method [36] is probably the most precise but requires steady-state concentrations of the analyte, and thus it has limited applicability in pharmacokinetic studies. Briefly, the method uses increasing concentrations of the analyte in the perfusion fluid. According to Equation 5.1, the concentration gradient is in the direction of the perfusate when the concentration in the interstitial fluid is larger than that in the perfusate, and in the other direction when the concentration of the perfusate is higher than that in the interstitial fluid. When the concentrations in the two fluids are the same, no net flux will be observed. That is, the concentration in the perfusate will not change. This method is mostly used for the titration of endogenous substances in the skin, such as histamine [37], glucose [38], lactate, and piruvate [39]. It can be used for exogenous substances as well, provided steady-state concentrations are obtained, for example, by continuous IV infusion. This method, however, requires a long execution time and it is thus difficult to apply. A "dynamic-no-net-flux" method was proposed [20,40] to overcome this deficiency but it requires several subjects and, again, is of little practical use in skin pharmacokinetics.

In the retrodialysis method [41] the analyte is added to the perfusate and its in vivo loss is calculated as:

$$Loss = \frac{[C]_{perfustate} - [C]_{dialysate}}{[C]_{perfustate}} \qquad (5.3)$$

The "loss" represents the fraction of analyte that leaves the perfusate. Theoretically, the factors determining loss (Equation 5.3) are the same as those that affect recovery (or gain, Equation 5.2); thus, loss and gain must be identical [6]. This method has been validated against the no-flux method [9,42] and it is considered a one-point version of the no-flux method [20]. It requires only one or two hours and can be performed using the same probe before or after an experiment. It can also be applied to an identical probe under the same experimental conditions when the tissue is large enough to permit the insertion of an additional probe without interfering with the primary experimental probe. This is particularly useful for studies of transdermal delivery in the skin.

Finally, there is the internal standard method [43] that combines the retrodialysis method with the use of a calibrator (or internal standard) added to the perfusate during the experiment. Changes in the recovery of the calibrator suggest changes in the recovery of the analyte and can be used to correct it. The main difficulty of this approach is finding a suitable calibrator that is chemically similar to the analyte but does not interfere with its pharmacokinetics in vivo. However, this method is conveniently performed during the experiment and also provides quality control data.

Simonsen et al. [44] compared the retrodialysis and the internal standard methods as applied to the estimation of true unbound extracellular concentrations in the dermis of hairless rats after topical delivery of C^{14}-salicylic compounds. The retrodialysis recovery was determined in a separate set of experiments, whereas the calibration method was performed for the individual probes during the experiments. The internal standard was H^3-salicylic acid. It was separately demonstrated that the two compounds have identical recovery characteristics and therefore H^3-salicylic acid is ideal for calibration. The two correction methods performed alike and the internal standard method did not lower the variability, contrary to expectation.

In conclusion, the choice of which method to use for in vivo probe calibration is mostly based upon the general scope of the project. The retrodialysis method is now the de facto standard method used in cutaneous pharmacokinetic studies.

5.4 INSERTION TRAUMA AND PHYSIOLOGICAL CONSIDERATIONS

Two major factors must be considered when microdialysis is used in vivo:

1. The initial tissue damage inflicted during the probe insertion procedure
2. The subsequent reaction of the tissue to the presence of the probe that may alter the performance of the probe during the course of the experiments

Anderson et al. [45] studied the effect on blood flow of a concentric-type (Y-type) probe insertion in the forearm of human volunteers with Laser Doppler perfusion imaging prior to, during, and after microdialysis probe insertion. They observed that probe insertion caused an increase in skin blood perfusion in the whole test area that subsided about 15 min after probe insertion. The circulatory response centered on the site of insertion and the tip of the probe. Skin perfusion levels returned to near normal values within 60 min. The authors concluded that 40 min is a reasonable time to allow for blood flow normalization before beginning the actual experiment. Krogstad et al. [46] performed biopsies on the forearm and abdominal skin of healthy volunteers around the area of the probe insertion to assess the structural impairment after 8–10 h of microdialysis. They observed small traces of local bleeding, but no evidence of edema or pathological extravasation of leucocytes that would indicate inflammation.

These studies were conducted in a few human subjects without local anesthesia. Even though the pain of a probe insertion is minimal and comparable to the insertion of a needle for IV administration, some investigators considered the use of local anesthesia to make the subjects more comfortable or simply to satisfy the IRB requirements. Typical local anesthesia is achieved with the administration of a local anesthetic (e.g., lidocaine, in the form of cream), or with ice packs. Benfeldt and Groth [14] studied the effects of local anesthesia (Xylocain, 10 mg/mL) on the insertion trauma in human forearms. They measured skin blood flow and erythema with Laser Doppler perfusion imaging and Dermaspectrometer. The trauma-induced edema and the effects of skin thickness were studied with ultrasound imaging. They found that local anesthesia prior to the insertion reduced the effects of trauma. Probe depth in the dermis did not influence the effects of trauma. They concluded that at least 90–120 min are required after insertion in order to allow the vascular reaction to needle trauma to return within the baseline range. The efficacy of the EMLA© cream (2.5% w/w lidocaine and 2.5% w/w prilocaine) to relive the pain of needle insertion was studied 20, 40, and 60 min after cream application [47]. They found that 40 min after application the pain was reduced by 75%.

Ice packs have an advantage because they can be applied on the area just 5 min before the insertion to induce local anesthesia [12].

When cutaneous microdialysis is performed in animals, total anesthesia or tranquilization is used rather than local anesthesia. Benfeldt and Groth [14] studied the effect of total anesthesia on basal skin blood flow and the trauma in the skin after insertion of a microdialysis probe in hairless rats anaesthetized either with halothane or pentobarbital sodium. Basal skin blood flow and the effect of insertion of a microdialysis probe were measured by laser Doppler perfusion imaging. Trauma induced by histamine release was also investigated. Rats anaesthetized with pentobarbital sodium showed a stable skin perfusion in contrast to rats anaesthetized with halothane. A significant increase in blood flow was observed after insertion of the microdialysis probe in the dermis of rats anaesthetized with pentobarbital sodium, whereas no change in skin blood flow was observed in rats anaesthetized with halothane. Probe insertion caused histamine release in rats. The authors concluded that a minimum equilibration period of 30 min between probe insertion and the start of the experiment is recommended to let the initial trauma subside and stabilize. Mathy et al. [48] studied the trauma induced by insertion of a linear microdialysis probe in the subcutaneous and dermal tissue in a rat with noninvasive bioengineering methods (TEWL, Laser Doppler Velocimeter, Chromameter) as well as with histology. The results showed that the dermal and subcutaneous insertion of microdialysis probes did not change skin permeability, blood flow, or color. The probe depth did not influence the trauma. No significant physical damage was observed after probe insertion.

Once inserted, the probe is used for a period of time determined by the experimental design. With time, endogenous substances such as proteins or lymphocytes, may infiltrate the membrane and the surrounding tissue and may alter the membrane operating parameters. This process is also known as *fouling* [11]. Different types of membranes may bind preferentially to different types of compounds. As reported above in the "Membrane" section lipophilic polymeric membranes tend to bind more to proteins and peptides than hydrophilic membranes. Ault et al. [49] investigated the skin response to probe implantation and perfusion in anesthetized fuzzy rats. Animals were sacrificed at 0.1, 6, 12, 24, 30, and 72 h and the skin around the probe underwent histological examination. Immediately after probe implantation there was no evidence of edema or substantial tissue disruption. At 6 h, lymphocytes started to infiltrate around the probe and increased during the subsequent period. Lymphocytes adhered to the surface of the membrane starting at 24 h. Retrodialysis studies performed in another set of rats showed that recovery was constant for the first 24 h, then a sharp increase occurred at 30 h and beyond, in coincidence with the formation of scar tissue around the dialysis membrane.

Similar studies in humans are not available (to the best of my knowledge). However, a 24 h sampling period is usually considered safe and it has become the de facto standard practice [50].

5.5 PROBE DEPTH

The knowledge of the exact location of the probe in the tissue is essential for a correct interpretation of the microdialysis experiments. In cutaneous microdialysis, the distance of the probe from the surface of the skin is called probe depth and provides an indication of the layer of the skin in which the probe is located. In studies of topically applied formulations, it is also a measure of the path length that a molecule must traverse before being collected by the probe. For this reason, the effect of probe depth on microdialysis data has been extensively studied. Probe depth can be determined noninvasively with ultrasound measurements both in humans and in animals. It has been demonstrated that the insertion trauma itself causes edema and may increase probe depth [14]. Intuitively, it would appear that the probe depth is a source of experimental variability. In fact, Muller et al. [51] found a significant correlation of the depth of the microdialysis probe with the area under nicotine concentration versus the time curve (AUC) and with the steady-state concentrations. However, in Muller's study, the probe depth ranged from 1.8 (dermis) to 4.5 mm (subcutaneous fat). Indeed, these findings were not confirmed in other studies in which the probe depth range was carefully maintained within the dermis range [12,44,52].

5.6 CONCLUSIONS

This chapter reviewed the basic concepts and practical aspects of cutaneous microdialysis. At this point, the reader should be able to design and perform successfully experiments to study the pharmacokinetics of drugs in skin.

REFERENCES

1. L. Bito, H. Davson, E. Levin, M. Murray, N. Snider, The concentrations of free amino acids and other electrolytes in cerebrospinal fluid, in vivo dialysate of brain, and blood plasma of the dog. *J Neurochem* 13 (1966) 1057–1067.
2. L. Groth, Cutaneous microdialysis. A new technique for the assessment of skin penetration. *Curr Probl Dermatol* 26 (1998) 90–98.
3. K. M. Steele and C. E. Lunte, Microdialysis sampling coupled to on-line microbore liquid chromatography for pharmacokinetic studies. *J Pharm Biomed Anal* 13(2) (1995) 149–154.
4. P. Sinko, *Martin's Physical Pharmacy and Pharmaceutical Sciences*. Lippincott Williams & Wilkins, Baltimore, MD, 2006.
5. G. Amberg and N. Lindefors, Intracerebral microdialysis: II. Mathematical studies of diffusion kinetics. *J Pharmacol Methods* 22(3) (1989) 157–183.
6. P. M. Bungay, P. F. Morrison, R. L. Dedrick, Steady-state theory for quantitative microdialysis of solutes and water in vivo and in vitro. *Life Sci* 46(2) (1990) 105–119.
7. L. Groth, Cutaneous microdialysis. Methodology and validation. *Acta Derm Venereol Suppl (Stockh)* 197 (1996) 1–61.
8. J. K. Hsiao, B. A. Ball, P. F. Morrison, I. N. Mefford, P. M. Bungay, Effects of different semipermeable membranes on in vitro and in vivo performance of microdialysis probes. *J Neurochem* 54(4) (1990) 1449–1452.
9. K. Hamrin, H. Rosdahl, U. Ungerstedt, J. Henriksson, Microdialysis in human skeletal muscle: Effects of adding a colloid to the perfusate. *J Appl Physiol* 92(1) (2002) 385–393.
10. H. Benveniste, Brain microdialysis. *J Neurochem* 52(6) (1989) 1667–1679.
11. J. A. Stenken, Methods and issues in microdialysis calibration. *Anal Chim Acta* 379 (1999) 337–358.
12. R. N. Tettey-Amlalo, I. Kanfer, M. F. Skinner, E. Benfeldt, R. K. Verbeeck, Application of dermal microdialysis for the evaluation of bioequivalence of a ketoprofen topical gel. *Eur J Pharm Sci* 36(2–3) (2009) 219–225.

13. F. X. Mathy, V. Preat, R. K. Verbeeck, Validation of subcutaneous microdialysis sampling for pharmacokinetic studies of flurbiprofen in the rat. *J Pharm Sci* 90(11) (2001) 1897–1906.

14. E. Benfeldt and L. Groth, Feasibility of measuring lipophilic or protein-bound drugs in the dermis by in vivo microdialysis after topical or systemic drug administration. *Acta Derm Venereol* 78(4) (1998) 274–278.

15. C. Carneheim and L. Stahle, Microdialysis of lipophilic compounds: A methodological study. *Pharmacol Toxicol* 69(5) (1991) 378–380.

16. M. Muller, R. Schmid, O. Wagner, B. V. Osten, H. Shayganfar, H. G. Eichler, In vivo characterization of transdermal drug transport by microdialysis. *J Control Release* 37(1–2) (1995) 49–57.

17. A. N. Khramov and J. A. Stenken, Enhanced microdialysis recovery of some tricyclic antidepressants and structurally related drugs by cyclodextrin-mediated transport. *Analyst* 124(7) (1999) 1027–1033.

18. K. W. Ward, S. J. Medina, S. T. Portelli, K. M. Mahar Doan, M. D. Spengler, M. M. Ben, D. Lundberg, M. A. Levy, E. P. Chen, Enhancement of in vitro and in vivo microdialysis recovery of SB-265123 using Intralipid and Encapsin as perfusates. *Biopharm Drug Dispos* 24(1) (2003) 17–25.

19. J. Kehr, A survey on quantitative microdialysis: Theoretical models and practical implications. *J Neurosci Methods* 48(3) (1993) 251–261.

20. E. C. de Lange, A. G. de Boer, D. D. Breimer, Methodological issues in microdialysis sampling for pharmacokinetic studies. *Adv Drug Deliv Rev* 45(2–3) (2000) 125–148.

21. K. V. Peinemann and S. Pereira Nunes, *Membrane for the Life Sciences*, Wiley-WCH, Verlag GmbH, 2008.

22. C. K. Poh, J. Lu, W. R. Clark, D. Gao, Kidneys, Artificial. In: G. Wnek and G. L. Bowlin (Eds.), *Encyclopedia of Biomaterials and Biomedical Engeneering*, Vol. 3, Informa Healthcare, Inc, New York, 2008, p. 1576.

23. Mulder, M., Basic Principles of Membrane Technology, 2nd edn., Dordrecht, the Netherlands: Kluwer Academic Publishers, 1996.

24. A. Kim, L. A. Suecof, C. A. Sutherland, L. Gao, J. L. Kuti, D. P. Nicolau, In vivo microdialysis study of the penetration of daptomycin into soft tissues in diabetic versus healthy volunteers. *Antimicrob Agents Chemother* 52(11) (2008) 3941–3946.

25. A. Klimowicz, S. Bielecka-Grzela, L. Groth, E. Benfeldt, Use of an intraluminal guide wire in linear microdialysis probes: Effect on recovery? *Skin Res Technol* 10(2) (2004) 104–108.

26. K. N. Fattah, Evaluation of an iontophoretic technique for the delivery of pentamidine to the skin, Master's thesis, *Division of Pharmaceutical Sciences*, Long Island University, Brooklyn, NY, 2003.

27. N. Solanki, *Pharmacokinetics of Triamcinolone Acetonide in Plasma and Skin Following IV-Infusion Administration in a Rabbit Model*, Master's thesis, *Division of Pharmaceutical Sciences*, Long Island University, Brooklyn, NY, 2009.

28. E. Camel, M. O'Connell, B. Sage, M. Gross, H. Maibach, The effect of saline iontophoresis on skin integrity in human volunteers. I. Methodology and reproducibility. *Fundam Appl Toxicol* 32(2) (1996) 168–178.

29. N. A. Monteiro-Riviere, Altered epidermal morphology secondary to lidocaine iontophoresis: In vivo and in vitro studies in porcine skin. *Fundam Appl Toxicol* 15(1) (1990) 174–185.

30. N. K. Mize, M. Buttery, P. Daddona, C. Morales, M. Cormier, Reverse iontophoresis: Monitoring prostaglandin E2 associated with cutaneous inflammation in vivo. *Exp Dermatol* 6(6) (1997) 298–302.

31. M. N. Berliner, Skin microcirculation during tapwater iontophoresis in humans: Cathode stimulates more than anode. *Microvasc Res* 54(1) (1997) 74–80.

32. H. Molitor and L. Fernandez, Studies on Iontophoresis. I. Experimental studies on the cause and prevention of iontophoretic burns. *Am J Med Sci* 198 (1939) 778–784.
33. P. C. Johnson and M. Intaglietta, Contributions of pressure and flow sensitivity to auto-regulation in mesenteric arterioles. *Am J Physiol* 231(6) (1976) 1686–1698.
34. G. Stagni, D. O'Donnell, Y. J. Liu, D. L. Kellogg, Jr., A. M. Shepherd, Iontophoretic current and intradermal microdialysis recovery in humans. *J Pharmacol Toxicol Methods* 41(1) (1999) 49–54.
35. H. J. Holovics, C. R. Anderson, B. S. Levine, H. W. Hui, C. E. Lunte, Investigation of drug delivery by iontophoresis in a surgical wound utilizing microdialysis. *Pharm Res* 25(8) (2008) 1762–1770.
36. P. Lonnroth, P. A. Jansson, U. Smith, A microdialysis method allowing characterization of intercellular water space in humans. *Am J Physiol* 253(2 Pt 1) (1987) E228–E231.
37. L. J. Petersen, Quantitative measurement of extracellular histamine concentrations in intact human skin in vivo by the microdialysis technique: Methodological aspects. *Allergy* 52(5) (1997) 547–555.
38. L. J. Petersen, J. K. Kristensen, J. Bulow, Microdialysis of the interstitial water space in human skin in vivo: Quantitative measurement of cutaneous glucose concentrations. *J Invest Dermatol* 99(3) (1992) 357–360.
39. A. L. Krogstad, P. A. Jansson, P. Gisslen, P. Lonnroth, Microdialysis methodology for the measurement of dermal interstitial fluid in humans. *Br J Dermatol* 134(6) (1996) 1005–1012.
40. R. J. Olson and J. B. Justice, Jr., Quantitative microdialysis under transient conditions. *Anal Chem* 65(8) (1993) 1017–1022.
41. L. Stahle, S. Segersvard, U. Ungerstedt, A comparison between three methods for esti-mation of extracellular concentrations of exogenous and endogenous compounds by microdialysis. *J Pharmacol Methods* 25(1) (1991) 41–52.
42. Y. Wang, S. L. Wong, R. J. Sawchuk, Microdialysis calibration using retrodialysis and zero-net flux: Application to a study of the distribution of zidovudine to rabbit cerebro-spinal fluid and thalamus. *Pharm Res* 10(10) (1993) 1411–1419.
43. M. R. Bouw and M. Hammarlund-Udenaes, Methodological aspects of the use of a cali-brator in in vivo microdialysis-further development of the retrodialysis method. *Pharm Res* 15(11) (1998) 1673–1679.
44. L. Simonsen, A. Jorgensen, E. Benfeldt, L. Groth, Differentiated in vivo skin penetra-tion of salicylic compounds in hairless rats measured by cutaneous microdialysis. *Eur J Pharm Sci* 21(2–3) (2004) 379–388.
45. C. Anderson, T. Andersson, K. Wardell, Changes in skin circulation after insertion of a microdialysis probe visualized by laser Doppler perfusion imaging. *J Invest Dermatol* 102(5) (1994) 807–811.
46. P. A. Jansson, A. L. Krogstad, P. Lonnroth, Microdialysis measurements in skin: Evidence for significant lactate release in healthy humans. *Am J Physiol* 271(1 Pt 1) (1996) E138–E142.
47. J. L. Cracowski, S. Lorenzo, C. T. Minson, Effects of local anaesthesia on subdermal needle insertion pain and subsequent tests of microvascular function in human. *Eur J Pharmacol* 559(2–3) (2007) 150–154.
48. F. X. Mathy, A. R. Denet, B. Vroman, P. Clarys, A. Barel, R. K. Verbeeck, V. Preat, In vivo tolerance assessment of skin after insertion of subcutaneous and cutaneous micro-dialysis probes in the rat. *Skin Pharmacol Appl Skin Physiol* 16(1) (2003) 18–27.
49. J. M. Ault, C. M. Riley, N. M. Meltzer, C. E. Lunte, Dermal microdialysis sampling in vivo. *Pharm Res* 11(11) (1994) 1631–1639.
50. P. Dehghanyar, B. X. Mayer, K. Namiranian, H. Mascher, M. Muller, M. Brunner, Topical skin penetration of diclofenac after single- and multiple-dose application. *Int J Clin Pharmacol Ther* 42(7) (2004) 353–359.

51. M. Muller, R. Schmid, O. Wagner, B. v. Osten, H. Shayganfar, H. G. Eichler, In vivo characterization of transdermal drug transport by microdialysis. *J Control Release* 37 (1995) 49–57.

52. G. Stagni, D. O'Donnell, Y. J. Liu, D. L. Kellogg, T. Morgan, A. M. Shepherd, Intradermal microdialysis: Kinetics of iontophoretically delivered propranolol in forearm dermis. *J Control Release* 63(3) (2000) 331–339.

53. J. A. Stenken, E. M. Topp, M. Z. Southard, C. E. Lunte, Examination of microdialysis sampling in a well-characterized hydrodynamic system. *Anal Chem* 65(17) (1993) 2324–2328.

54. L. Sun, J. A. Stenken, Improving microdialysis extraction efficiency of lipophilic eicosanoids. *J Pharm Biomed Anal* 33(5) (2003) 1059–1071.

55. J. A. Stenken, R. Chen, X. Yuan, Influence of geometry and equilibrium chemistry on relative recovery during enhanced microdialysis. *Anal Chim Acta* 436(1) (2001) 21–29.

6 Sampling Substrates by Skin Permeabilization

M. Begoña Delgado-Charro

CONTENTS

6.1 INTRODUCTION TO TRANSCUTANEOUS SAMPLING OF DRUGS

Noninvasive sampling methodologies offer obvious benefits to all patients: (a) frequent sampling provides better information, (b) reduced pain and discomfort (i.e., better compliance), (c) low risk of infection, and (d) potential for home monitoring.

Furthermore, noninvasive diagnosis and monitoring would be particularly useful for patients who are repetitively subjected to invasive blood withdrawal procedures to perform drug monitoring, especially the "critically" ill, the elderly, and the pediatric patients, for whom the need for noninvasive diagnosis and monitoring is particularly acute. In the long term, one would expect that the development of noninvasive techniques that measure drug concentrations in body fluids would allow determining drug pharmacokinetics in poorly characterized populations.

The skin offers a large and easily accessible surface for drug sampling. Transdermal passive diffusion, the first and simplest approach, is severely limited by the skin's outermost layer, the stratum corneum (SC), which renders passive extraction of many drugs too slow for practical applications. Nevertheless, passive sampling brought up knowledge to the field and identified the key issues, which future technologies would have to resolve.

The concept of continuous transepidermal drug collection (CTDC) as a tool to determine drug intake and pharmacokinetics was proposed by Peck et al. [1] in the early 1980s. A series of models combining pharmacokinetics with extraction kinetics were developed assuming first-order transfer processes between the systemic and the collector compartments. A series of simulated profiles indicated the key requirements for the CTDC of a given drug to be feasible: the transfer back from the collector to the central compartment must be low, and the duration of the CTDC relative to the elimination half-life of the drug. Further, the constant of rate transfer from the central to the collector compartment must be high relative to the other processes, and frequent, precise, and accurate measurement of the amount collected at different times is necessary. The outward transdermal migration of theophylline into gel/activated carbon collectors in fuzzy rats was first demonstrated by Peck et al. [2] and subsequently, similar experiments were conducted in rhesus monkeys and humans [3,4]. The results confirmed the validity of the expression $Q = Kp \cdot A \cdot \int_{t_2}^{t_1} (C_p - C_{tcs}) \, dt$, where Q is the cumulative amount collected during the time interval $(t_1 - t_2)$, K_p is the apparent permeability coefficient, A is the area of contact, and C_p and C_{tcs} are the chemical concentrations in the cutaneous capillary plasma and in the transcutaneous collector medium, respectively. It can be considered that C_{tcs} is negligible and therefore, $Q = Kp \cdot A \cdot \mathrm{AUC}_{t_1}^{t_2}$. The significant effects of occlusion and of the collecting formulation on the transdermal collection of theophylline, methotrexate, and parathion were later demonstrated in vitro and in vivo in rats [5]. Theophylline and caffeine were also collected in preterm children (2–89 days old) by means of four transcutaneous collector systems attached to the back or the abdomen for 4–12 h [6]. While there was a positive relationship between the amounts collected and plasma concentrations, the amounts were poor predictors of individual plasma concentration. Surprisingly, the effluxes increased with postconceptional age. Caffeine was collected in healthy volunteers (Figure 6.1), and the significant effects of heat and possibly sweat on CTDC were demonstrated [7].

This early series of experiments clearly proved that sampling noninvasively across the skin was feasible and that the collected amounts were positively related to systemic drug exposure. However, they also identified key areas for development before the technique could be applied efficiently. A large inter- and intra-variability that could be caused by (a) the device (poor skin contact, vehicle effects, and occlusion effects), (b) analytical challenges (extraction from and assay of low amounts of drug in the collecting matrix), and (c) effects of temperature, environmental conditions, and exercise on capillary blood flow, skin permeability, and sweat contribution. A significant challenge was the discrimination between changes in fluxes caused by transdermal flux modifications and those caused by the drug pharmacokinetics, for example, an increment in the collected amount could be due to both a higher systemic exposure or to an occlusion effect. Passive diffusion was too slow and unpractical for drugs having low passive permeability, as sampling periods were extended to allow the collection of sufficient drug to be accurately measured. Thus, a faster mechanism of extraction less prone to intra- and inter-variability and

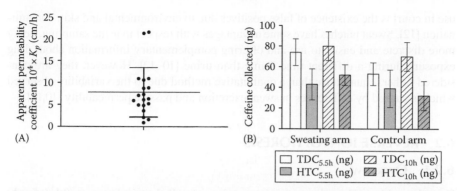

FIGURE 6.1 Passive transcutaneous collection of caffeine in healthy subjects. (A) Apparent in vivo permeability coefficient (K_p) determined for 19 subjects. K_p was obtained from the slope ($\beta = K_p \cdot A$) of the regression line $Q = \beta \cdot AUC$, where Q is the amount of caffeine collected in each transcutaneous collection device (TCD), AUC is the area under the plasma concentration–time curve from the time of the first dose until the time of TDC removal, and A is the skin contact area. Each point represents an individual value; the lines correspond to the mean ± SD. (B) Effect of heat application to the arm ("sweating") on caffeine collection into TDC and Hill Top chamber (HTC) in 14 subjects. The contra lateral arm of the same individual was the "non-sweating" control. Time dependencies and a large variability was observed. Data given as mean ± SEM. For a detailed explanation on this studies, see Ref. [7]. (Data redrawn from Conner, D.P. et al., *J. Invest. Dermatol.*, 96, 186, 1991.)

which could expand the range of drugs available for extraction was required. Finally, these experiments required plasma concentration—AUC values that combined with the collected amounts allowed estimation of the in vivo permeability coefficient. Unfortunately, these values were quite variable (Figure 6.1), meaning that individual calibration would be necessary. It follows that either an extraction method that does not need calibration, or a calibration method that requires none or just a nominal blood sample, should be developed for CTDC to become truly noninvasive.

Later on, Pellett et al. [8] investigated glucose and titrated water back diffusion in vitro. A series of formulations that modified osmotic pressure were tried; however, transport was only significantly enhanced when the skin was tape stripped. It was concluded that passive diffusion on its own, is not a sufficiently fast and efficient way of sampling chemicals across the skin, unless the stratum corneum is permeabilized to some degree.

Thus, the idea of passive CTDC for drugs and glucose monitoring and pharmacokinetic profiling was discarded in favour of more efficient sampling methods. Nevertheless, sweat patches such as Pharm-check™ commercialized by Pharmchem™ labs have been developed for drug of abuse testing, and have found application in monitoring and probation programs. Despite the name, it is expected that drug accumulation into the patch occurs by a combination of sweat excretion and passive diffusion. The Pharm-check patch has a ~14 cm² surface, is non-occlusive, and collects a minimum 300 μL of sweat per day. The patch is worn for a week, after which it is removed and sent for analysis. It has been used for benzodiazepins, buprenorphine, codeine, phenobarbital, heroin, cocaine, and their metabolites [9–15]. A concern regarding their use in court is the existence of false positives due to environmental and skin contamination [12]. Sweat patches have some advantages with respect to urine sampling, being more discrete and easier to use and offering complementary information about drug exposure during a different time frame than urine [10–12]. However, they are considered rather a qualitative than a quantitative method due to the variability observed which is caused by differences in sweat excretion and passive permeability [10,12].

6.2 REVERSE IONTOPHORESIS

6.2.1 OVERVIEW

Transdermal iontophoresis consists in applying a small electrical current (0.1–0.5 mA) to the skin in order to increase percutaneous molecular transport. This technique has found applications in the field of transdermal drug delivery, and several iontophoretic drug delivery systems have reached the market. Several reviews on the mechanisms of transport underlying iontophoresis and its applications for drug delivery are available elsewhere [16–20]. Iontophoresis is considered as a symmetric process, which means that molecules from within the subdermal compartment can be extracted to the skin surface. Soon, its potential for noninvasive clinical chemistry was obvious. Applications include general blood chemistry, glucose monitoring, the detection of diagnostic markers, and therapeutic drug monitoring [21].

Briefly, both electromigration and electroosmosis contribute to iontophoretic transport, allowing sampling of both charged and neutral analytes. Faraday's law can be used to relate the ionic transport (either inward or outward) crossing the membrane to the intensity of the electric current applied, the time of current passage, and the charge per ion: $M_i = T \cdot I \cdot t_i / F \cdot z_i$ where M_i is the number of moles of the "i" ion flowing through the skin in a time $T(s)$, z_i is the valence of ion "i", F is Faraday's constant, I is the intensity of the current applied, and t_i is the transport number of the ion "i" [19,21]. It is important to realize that all the ions present at both sides of the skin compete to transport charge during iontophoresis.

Thus, the iontophoretic extraction of an analyte will be determined by

1. The time of each sampling period (T); this is usually determined by analytical criteria, as sufficient drug has to be extracted during the period T to allow precise quantification. Extended sampling periods facilitate quantification but then iontophoresis provides information of the average level of the analyte in the body during the sampling period considered.
2. The intensity of current that is directly and easily controlled by the power supply and is usually fixed taking into account the area of application. For example, the density of current should not exceed the value of 0.5 mA/cm² [22].
3. The valence of the extracted ion is dictated by its molecular structure and the pH of the surrounding fluids.
4. The transport number of the ion of interest. Transport numbers are difficult to predict but it is known that in the presence of competing ions, as is the case of reverse iontophoresis, the transport number of an ion will be directly proportional to its concentration in the subdermal fluid and to its

electrical mobility [19,23–27]. Obviously, the ions that are more mobile and present in high concentrations, will play the main role in the movement of charge through the skin. This means that sodium and chloride ions will transport most of the charge towards the cathodal and anodal compartment, respectively, as has been experimentally shown [28–30].

While experimental conditions can be easily optimized in the case of iontophoretic drug delivery (e.g., by minimizing the presence of competing species and increasing the drug concentration and its ionized fraction), this manipulation is not possible in reverse iontophoresis. The concentration of a drug will be given by the dosage regimen and the relevant pharmacokinetics, and that of endogenous substances will correspond to either the normal or pathological ranges. Furthermore, only the ionized fraction of the analyte is extractable by electromigration and the value of this fraction will depend on the relevant pK_a and the physiological pH. In the case of analytes subject to protein and tissue binding, only the free fraction can be significantly extracted across the skin [31]. Thus, reverse iontophoresis provides measurements of free drug concentration, an interesting feature for some drugs (Figure 6.2). Finally, the ionic mobility (which contributes to the value of the transport number) is inversely related to molecular size [23]. In summary, the best candidates to be

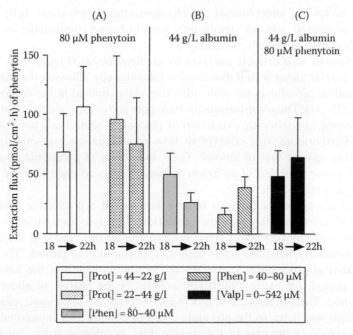

FIGURE 6.2 Monitoring of free phenytoin by reverse iontophoresis [31]: sum of anodal and cathodal extraction fluxes of phenytoin (mean ± SD, $n \geq 6$) in response to changes in the free fraction of the drug in the subdermal solution. The changes described were made immediately after the measurement taken at 18 h. (A) Impact of changing albumin concentration. (B) Impact of changing phenytoin total concentration. (C) Addition of a competing drug, valproate. (Data redrawn from Leboulanger, B. et al., *Eur. J. Pharm. Sci.*, 22, 427, 2004.)

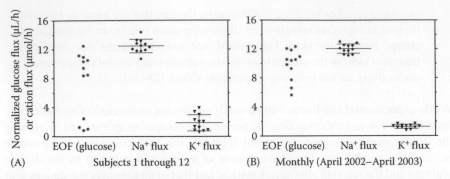

FIGURE 6.3 Glucose, sodium, and potassium extraction by reverse iontophoresis in healthy volunteers [29]. Normalized glucose fluxes correspond to the ratio: glucose flux/glucose blood concentration. Each individual data point corresponds to the mean of 10 intervals of extraction of 15 min. Sodium extraction was essentially constant. The solid lines show the mean ± SD for the sodium and potassium data sets. (A) Inter-subject variability in 12 volunteers. The normalized glucose fluxes for 4 volunteers were significantly lower than those for the rest. (B) Intra-subject variability in one subject as a function of time over a calendar year. Normalized glucose fluxes were significantly lower during the winter months. (Data redrawn from Sieg, A. et al., *Clin. Chem.*, 50, 1383, 2004.)

extracted by reverse iontophoresis (by electromigration) are small, fully charged, highly concentrated, and not significantly protein-bound compounds; an excellent case has been that of lithium, for example.

Iontophoresis also extracts analytes by electroosmosis (Figure 6.3). This solvent flow carries along with it dissolved solutes thereby allowing the transport of neutral, and, especially, polar molecules that are extracted at the cathodal compartment [32–34]. The electroosmostic transport is the basic mechanism underlying the reverse iontophoretic extraction of glucose, a neutral and polar substance [35–37]. Furthermore, this convective flow reinforces the transport of cations while acting against that of anions. Thus, extraction of a cationic analyte will always be easier than that of an anion of similar physico-chemical and pharmacokinetic properties [38].

Reverse iontophoresis is much more efficient and reproducible than passive transdermal extraction. Even so, the quantities of analyte transported to the skin surface are typically very small. The concentration of the analyte at the collection chamber is usually 10–1000 lower than its concentration in plasma. The sensitivity of analytical procedures determines the minimum length of the iontophoretic sampling period that extracts sufficient amount of the analyte to allow accurate quantification. Thus, in the case of glucose monitoring, an impressive refinement of the technique was done, so the glycemia could be continuously monitored [39–45]. On the contrary, in the case of therapeutic drug monitoring at the "steady state" the problem is less important (i.e., prolonged sampling times to obtain an average concentration could be acceptable to assess drug exposure). For example, an iontophoretic patch could be worn for a few hours at home, and then sent to the clinical chemistry laboratory. The "off-line" analysis and quantification would require a sampling device simpler and less expensive to design.

It should be kept in mind that transdermal iontophoresis extracts compounds from the skin and not directly from the blood. Iontophoretic fluxes of extraction are proportional to the analyte concentration in the subdermal interstitial fluid, and their information about plasma levels will depend on the rapidity with which the equilibrium plasma–interstitial fluid levels is established. Glucose interstitial levels are typically (but consistently) delayed 20 min with respect to plasma levels [29,46]; on the contrary, subdermal and systemic lactactemia are not obviously correlated [30,47,48].

The skin may also accumulate the analyte of interest such that the initial extraction sample contains mostly information about this local "reservoir" (this is the case for glucose [36], lithium [28], urea [49–51], lactate [30], and many amino acids [52–54]). A "warm-up" period is necessary to empty this reservoir before the iontophoretic fluxes are reflective of systemic levels. On the other hand, this phenomenon could be considered as an opportunity to obtain additional information from iontophoretic data (Figure 6.4). Because this first short phase extracts drugs and endogenous compounds stored in the skin reservoir, it could be exploited to gather "historical" knowledge regarding treatment compliance and average levels of endogenous markers, markers of skin health or with cosmetic significance. Once the skin reservoir is exhausted, analytes are extracted from the interstitial subdermal fluid, which provides "real time" information.

Finally, reverse iontophoresis will simply not work for some molecules that are too large to be extracted in quantifiable amounts in meaningful times.

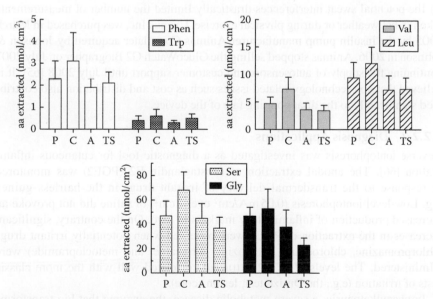

FIGURE 6.4 Extraction of amino acids from the skin reservoir by passive diffusion (P), iontophoretic extraction at the anode (A) and at the cathode (C), and by tape stripping (TS). The data correspond to the mean ± SD of six experiments done with dermatomed pig skin. The subdermal solution was PBS. Six representative examples were selected form the extensive data set comprising thirteen amino acids and glucosereported by Sylvestre et al. (From Bouissou, C.C. et al., *Pharm. Res.*, 26(12), 2630, 2009.)

6.2.2 Reverse Iontophoresis Applications

6.2.2.1 Glucose Monitoring by Reverse Iontophoresis

The first objective attempted was the iontophoretic extraction of glucose, as such a noninvasive tool would be of immense medical benefit in the management of diabetes [55]. Proof of concept was soon established first in vitro [36] and then in vivo in nondiabetic subjects [37] and an integrated device, the GlucoWatch Biographer® [41–44,56–59], which tracks changes in the blood sugar levels of diabetics over the entire range of glycemia, was developed by Cignus, Inc. and commercialized. The GlucoWatch Biographer received a CE Mark in 1999 and 2000 for adults and children (7–17 years), respectively, and FDA approval in 2001 and 2002 for adults and children (7–17 years), respectively [41,56,60]. A highly optimized adaptation of the Pt-glucose oxidase sensor [61] was developed to measure the very small amount of glucose extracted during each sampling period: 20 min initially and 10 min in the improved GlucoWatch G2 Biographer [62]. Several studies investigated the accuracy of the system to predict hypoglycemia [43,62], the effect of exercise [63], and the possible pre-application of corticosteroids to reduce skin irritation [64]. A comprehensive list of trials and studies was compiled by Tura et al. [60]. The GlucoWatch Biographer provided the first truly noninvasive approach to the monitoring of blood glucose but some limitations were apparent: (a) the lengthy 2–3 h, warm-up time before measurements can be made (due to the need to empty a glucose reservoir in the skin), (b) the fact that a "finger-stick" blood measurement is essential to calibrate the device and (c) the potential sweat interferences drastically limited the number of measurements taken in hot weather or during physical exercise. Cygnus Inc. was purchased in March 2005 by the insulin pump manufacturer Animas Corp. later acquired by Johnson & Johnson in 2006. Animas stopped selling the GlucoWatch G2 Biographer on July 2007 continuing the supply of autosensors and customer support until July 2008 [65]. It is believed that non-technology-related issues such as cost and distribution also contributed significantly to the disappointing fate of the device.

6.2.2.2 Diagnosis Applications

Reverse iontophoresis was investigated as a diagnostic tool for cutaneous inflammation [66]. The anodal extraction of Prostaglandin E2 (PGE2) was monitored in response to the transdermal delivery of irritant drugs in the hairless guinea pig. Low-level iontophoresis ($0.05 \, mA/cm^2$ over $2 \, cm^2$) of saline did not provoke an increased production of inflammatory markers in vivo. On the contrary, significant increases in the extraction of PGE2 were observed when potentially irritant drugs (chlorpromazine, chloroquine, promazine, tetracaine, and metoclopramide) were administered. The levels of PGE2 extracted correlated well with the more classic tests of irritation (e.g., the Draize test, lesion score).

In phenylketonuria, a severe metabolic disease, the enzyme that bio transforms phenylalanine is missing; the resulting continuous accumulation of this amino acid has a tragic impact on the development of children suffering from this disease [67]. Early detection of phenylketonuria and subsequent control of the diet are therefore essential. Thus, a noninvasive method of sampling phenylalanine in the paediatric population would be desirable. The amounts of the zwitterionic phenylalanine extracted at the cathode were proportional to the subdermal concentrations in early in vitro [34] and in vivo [68] studies. Subsequent in vitro and in vivo studies demonstrated that most amino acids could be extracted by reverse iontophoresis and quantified by LC-MS [52,53] or ion chromatography [54,69]. The main findings of these studies were (a) the existence of a vast skin reservoir for most amino acids (Figure 6.4), (b) zwitterionic aminoacids are more abundant in human than in porcine SC, (c) iontophoretic fluxes of extraction only report on systemic levels once the skin reservoir is exhausted, a step requiring several hours in humans in the case of some zwitterionic amino acids including phenylalanine, that is, too long for practical applications, and (d) zwitterionic amino acids were easily extracted by passive diffusion, cathodal and anodal iontophoresis although extraction towards the cathode was more efficient. These results have provided comprehensive information on the amino acids' abundance in the SC, and amino acids are part of the natural moisturizing factor and could possibly be used as markers of skin health.

The reverse iontophoretic extraction of urea by electroosmosis was performed in 17 patients (21–35 years) with impaired kidney function [70]. It was shown that the amounts of urea extractable pre- and post-dialysis were quite different in six pediatric patients (9–16 years). However, later research [46,49,51] showed that short-term reverse iontophoresis extracts an important amount of urea from the skin reservoir. Thus, the reduction in iontophoretic fluxes observed during dialysis could result as well from the progressive depletion of the skin reservoir. Wascotte et al. [49,71] optimized the procedure to monitor urea, potassium, and sodium in vitro and in vivo in healthy volunteers and patients with chronic kidney disease (CKD). It was shown that a good correlation between subdermal urea concentration and fluxes of extraction was obtained only after 1–2 h of iontophoresis, once the skin reservoir was depleted. It was also shown that 90 min iontophoresis could differentiate between healthy volunteers and patients with stage 5 CKD. However, unless the technique is further optimized, interpretation of iontophoretic fluxes of extraction during hemodialysis is complicated by the contribution of the urea reservoir in the skin. The development of noninvasive techniques to monitor renal function would result in an improved therapy and a better quality of life for the patient. In subsequent studies, Djabri et al. [72,73] studied the reverse iontophoretic extraction of iohexol, a marker of glomerular filtration rate. Iohexol has a relatively large molecular weight (821 Da) and is neutral and hydrophilic ($\log P = -3$), and thus it is only extracted by electroosmosis. Iontophoretic fluxes of iohexol followed the subdermal profiles of iohexol in one- and two-compartment model simulations in vitro. Reverse iontophoresis also provided, non-invasively, a reasonable estimate for the terminal rate constant in four paediatric patients [74].

Noninvasive lactatemia monitoring is of interest in the case of critically ill patients as well as a marker of performance in sports training [75,76]. Lactate, a relatively concentrated, small anion, constitutes a good candidate for reverse iontophoretic extraction. In fact, it is easily and rapidly extracted both in vitro and in vivo [30]. However, interpretation of the data obtained is extremely complicated by the extensive skin reservoir and the lactate's ubiquitous metabolic role; for example, subdermal interstitial lactate levels are not necessarily in equilibrium with systemic values [48].

6.2.2.3 Therapeutic Drug Monitoring

The use of reverse iontophoresis for noninvasive drug monitoring and pharmacokinetic profiling has been investigated for valproate [77], lithium [28,78,79], phenytoin [31], amikacin [80], caffeine, and theophilline [81]. Briefly, the data suggest that iontophoresis can efficiently and noninvasively sample drugs of appropriate physicochemical and pharmacokinetic properties, perform drug monitoring, and provide an estimation of the elimination (terminal) rate constant.

Lithium, a drug used to treat bipolar disorders, has been successfully monitored via reverse iontophoresis (Figure 6.5 and Table 6.1). First of all, Li^{+1} is a small, mobile,

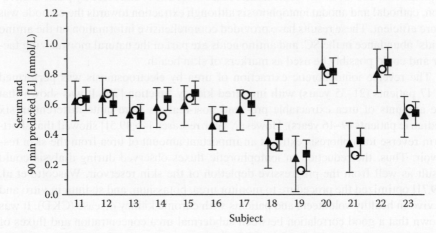

FIGURE 6.5 Therapeutic monitoring of lithium via reverse iontophoresis [28]. Comparison between measured C_{Li} (open circles) in a "test group" of 13 patients and the predictions (mean ± 95% confidence interval) based upon (A) J_{Li+} or iontophoretic extraction flux lithium (filled triangles) and (B) $R_{Na+} = J_{Li+}/J_{Na+}$ or lithium fluxes normalized using the internal standard sodium (filled squares). Values were determined using the equation $C_{Li} = C_{Li} \cdot R_{Na+}/K$ and K was determined in a different group of patients ($n = 10$) used as the "training set." (Data redrawn from Leboulanger, B. et al., *Clin. Chem.*, 50, 2091, 2004.)

TABLE 6.1
Determination of Pharmacokinetic Parameters
(K_e = Elimination Rate Constant and $t_{1/2}$
Elimination Half Life) via Reverse Iontophoresis
in a Simulated Bolus Experiment (Mean ± SD, $n = 6$)

Data	r^2	K_e (10^{-3} min^{-1})	$t_{1/2}$ (min)
Subdermal C_{Li}	0.99	4.2 ± 0.3	168 ± 14
J_{Li}	0.99	3.9 ± 0.2	180 ± 10
R_{Na} (J_{Li}/J_{Na})	1.00	3.8 ± 0.2	181 ± 9

Source: Data taken from Leboulanger, B. et al., *Pharm. Res.*, 21, 1214, 2004.

---

and nonprotein-bound cation that competes to transport charge across the skin much more efficiently than most therapeutic drugs. Second, plasma concentrations are high for lithium. As a result, the amounts detected at the skin surface can be easily assayed with existing technology. In vitro studies first demonstrated the linearity and rapidity of Li^+ extraction from a physiological buffer; in fact, when the concentration of Li^+ in the subdermal compartment was varied to simulate a pharmacokinetic profile, the extraction fluxes closely reproduced the contour [79]. In addition, the elimination rate constant and half-life could be directly, and noninvasively, estimated from the extraction flux data (Table 6.1). Subsequently, a study [28] with bipolar patients (chronically treated with lithium) showed the potential of reverse iontophoresis as a useful clinical tool for noninvasive monitoring of this drug (Figure 6.5). The reverse iontophoretic extraction fluxes were extremely well correlated with the corresponding plasma concentrations. Furthermore, the constant of extraction (the constant relating lithiemia with the iontophoretic extraction flux) determined with a subset of patients, could be successfully used to predict lithiemia in another group of subjects. A skin reservoir of lithium was found in these patients and was practically depleted after the first sampling time of 30 min. It would be worth exploring the use of this reservoir as an indication of exposure to therapeutic drugs (monitoring) and to abuse, doping, and harmful chemicals as it would provide a different time frame of exposure than urine, nails, and hair.

The potential of this noninvasive approach in the case of sick neonates has been explored in a model designed to mimic the developing cutaneous barrier of this population [81]. Studies with caffeine and theophylline demonstrated the feasibility of this approach for full-term infants, whose stratum corneum is fully developed. On the contrary, the technique is less satisfactory in the case of premature babies. In this case, the significant passive transport of the drug through the defective skin is superimposed upon the electrotransport of the target analyte, which confounds straightforward interpretation of the data.

Some studies have shown the impact of protein binding on the reverse iontophoretic extraction of highly protein-bound drugs such as the anti-epileptic valproic acid [77] and phenytoin [31]. It is expected that only the free drug will be transported across the skin as the protein-bound form is too large to be extracted. First, it was demonstrated that the reverse iontophoresis extraction of both drugs was concentration dependent over a wide range including free levels typically observed in patients. The negatively charged valproate was extracted to the anode, while phenytoin ($pK_a = 8.3$) was recovered at both anode (the ionized fraction of the drug being attracted by electromigration) and cathode (the neutral form being carried by electroosmosis). Following experiments showed that the amount of drug extracted was reduced consistently with a decreased free drug level caused by the presence of additional albumin in the subdermal solution. Conversely, displacement of phenytoin by valproate led to an increase in the amount of phenytoin sampled (Figure 6.2). These results support the potential usefulness of iontophoresis for monitoring free levels of drugs. On the other hand, the free concentrations of valproic acid and phenytoin are quite low (50–105 and 4–8 µm, respectively) and consequently, the amounts extracted by reverse iontophoresis are extremely small (the experiments described above used radiolabelled drugs). It follows that the analytical requirements for practical applications in vivo will be very demanding.

6.2.3 FUTURE PERSPECTIVES AND STRATEGIES FOR REVERSE IONTOPHORESIS OPTIMIZATION

Reverse iontophoresis offers a clear potential as a noninvasive sampling technique, yet there is a need to improve the existing technology and to optimize the interpretation of the information provided. A first issue is the calibration required to link fluxes of extraction with subdermal interstitial levels. The GlucoWatch Biographer required calibration via a conventional finger-stick glucose blood measurement. This invasive calibration essential to relate the amount of sugar extracted to the blood concentration was perceived as a disadvantage among the users of the device. During reverse iontophoresis, a certain amount of the analyte is recovered at the skin surface and diluted in a volume of "acceptor" fluid. The final concentration in the collector chamber depends on the efficiency of extraction and the volume of the "acceptor" solution. For example, during each sampling period, the GlucoWatch Biographer dilutes a flow of less than 1 µL into a volume of 400 µL [82], that is, a three order of magnitude dilution. Further, the electroosmotic glucose extraction, suffers from both inter- and intra-individual variability (Figure 6.3) [29].

Calibration of the GlucoWatch Biographer is required to correct for variation in the skin permeability to glucose and in the biosensor sensitivity [41]. The issue of calibration is even more critical in the case of therapeutic drug monitoring or pharmacokinetic studies when only one or just a few blood samples are taken.

To address this issue, the concept of an "internal standard" has been proposed [77]. Reverse iontophoresis simultaneously extracts numerous compounds; that is, the process is nonspecific, and, in fact, the specificity for a particular analyte (A) is only provided by the analytical assay. The internal standard (IS) is defined as a second substance (a) whose blood concentration is effectively constant and (b) is simultaneously extracted with the analyte, and then also analyzed in each sampling period. It follows that there should be a proportionality between the measured extraction flux ratio of the analyte and the internal standard, and the ratio of their subdermal concentrations. A critical step is the identification of the appropriate internal standard for each analyte. Initial proof-of-concept experiments were done in vitro, and the extraction of valproate [77] was normalized using glutamic acid as an anionic internal standard. It was shown that (a) the extraction flux of valproate varied linearly with its subdermal concentration, (b) the extraction flux of glutamate remained constant as the valproate concentration fluctuated, and (c) the ratio of the valproate to glutamate extraction fluxes was proportional both to their subdermal concentration ratio, and, as the glutamate concentration was fixed, to the valproate concentration. The approach reduced the experimental variability and permitted the subdermal valproate concentration to be predicted.

Similar success was achieved with the Li^+/Na^+ (A/IS) couple [28,79]. Sodium ions have been identified as the ideal internal standard to normalize the extraction of cations: (a) physiological sodium ion concentration does not vary outside the range of 125–145 mm (and typically remains within a much narrower window) and (b) Na^+ is the major charge carrier in iontophoresis in the outward direction towards the cathode. Because transport numbers are primarily determined by the mobility and the molar fraction of an ion, the sodium transport number is practically constant in the physiological range [25,83]. This was confirmed by in vivo measurements in patients and in healthy volunteers (Figure 6.3) [28,29]. Sodium was successfully used as an IS for lithium, and the calibration constant determined for a training-set of patients could be used to predict lithiemia in a testing-set of patients (Figure 6.5) [28]. However, while Na$^+$ was a valid IS for glucose in vitro [83], a subsequent in vivo study [29] revealed that electroosmotic flow is a much more sensitive phenomenon and can vary by nearly a factor of 10 even while the electromigrative flux of Na$^+$ remains unchanged (Figure 6.3). Subsequent work [46] has explored the possible use of urea (also electroosmotically extracted) as an IS for glucose. While urea behaved as a better IS for glucose than sodium, some variability remained. In summary, the validity of the internal standard technique has been demonstrated but the adequate internal standards have only been identified for lithium.

A second issue concerns the lack of models which integrate the kinetics of extraction with the subdermal pharmacokinetics of a drug. The use of reverse iontophoresis as a noninvasive sampling technique assumes the outward drug flux to be proportional to the drug subdermal concentration; in other words, $J_{drug} = \gamma \cdot [drug]_{subdermal}$, where γ is a constant. However, it has been shown [31,72,77,79] that this equation applies only after a so-called stabilization time. The value of γ increases in the initial period of extraction until it plateaus at a constant value, and during this stabilization time both changes in drug concentration and γ determine the magnitude of the fluxes of extraction. The length of this stabilization time is variable (~1–2 h for Li$^+$, 4–5 h for valproic acid, 3–4 h for iohexol, and >10 h for phenytoin) and is probably linked to the physicochemical (MW, log$P$, mobility, ionization) and pharmacokinetic (concentration, free fraction) properties of each drug. Thus, models need to be developed for pharmacokinetic profiling via reverse iontophoresis, which include the kinetics of the extraction, and, eventually, the contribution of a skin reservoir whenever relevant. Once these novel models are developed and validated, they will allow correct analysis of the data and determination of the pharmacokinetic parameters with the precision required for clinical purposes.

In summary, recent progress has demonstrated the considerable potential of reverse iontophoresis as a noninvasive tool in clinical chemistry and therapeutic drug monitoring. The GlucoWatch Biographer represented an important milestone for the technology as the first truly noninvasive monitoring device for diabetics. Reverse iontophoresis can be used to perform noninvasive therapeutic monitoring of well chosen drugs and improve the quality of care (and life) in patient populations for which repetitive blood sampling represents a significant burden. Pediatric, geriatric, and chronically ill individuals are obvious examples. However, the smaller size of the commercial markets (i.e., lithium monitoring relative to that for glucose monitoring) may discourage the ultimate realization of practical devices. Future applications will be largely determined by the evolution of analytical tools in terms of sensitivity, specificity, and miniaturization. Further understanding of (a) the type of information (the "historical" pool and the "real" time) provided by iontophoresis, (b) the kinetics of iontophoretic extraction, and (c) the calibration procedure is also required.

6.3 ELECTROPORATION-MEDIATED DRUG SAMPLING

Electroporation increases skin permeability by application of short (milli- to microseconds) electrical high voltage pulses which are believed to form transient aqueous pathways [84]. Extensive work on skin sampling via electroporation has been done by Murthy et al. who investigated the transcutaneous extraction of drugs (salicylic acid [85], acyclovir [86], ciprofloxacine [87], 8-methoxypsoralen [87], cephalexin [88]), and glucose [89]. In vivo studies were done on rats; briefly, the collection chamber was filled with 0.05% sodium dodecyl sulphate (SDS) in phosphate-buffered saline (PBS) and 10–15 pulses (1 ms, 120 V, 1 Hz) were applied. After this pretreatment, the SDS solution was replaced with PBS and a collection period of 15 min followed. In vitro and in vivo studies indicated the presence of SDS to reduce greatly the lag time in salicylic acid extraction and the pretreatment to increase permeability by ~two orders of magnitude [85]. Salicylic acid efflux started declining ~1 h after the treatment, probably due to skin recovery. Thus, the procedure had to be repeated for each time point in order to obtain a 6 h concentration-versus-time profile. A preliminary experiment was also necessary to estimate the calibration factor which allows relating the drug's efflux with its free drug concentration in the extracellular fluid ($C_{u,ECF}$). Once this factor was available, a good agreement was observed between the predicted $C_{u,ECF}$ and that determined by transdermal extraction. In subsequent work, acyclovir (Table 6.2) was administered via IV bolus to rats [86]; the dermal concentration of drug determined by microdialysis and electroporation-transdermal extraction was compared successfully. A subsequent study with glucose [89] used 30 pulses (1 ms, 120 V/cm^2) in the absence of SDS, and a good correlation ($r^2 = 0.87$) was observed between the rats' venous concentration and the glucose extracted transdermally. Finally, other studies, also in vivo in rats, applied 30 electrical pulses (10 ms, 120 V/cm^2, 1 Hz) in the absence of SDS to extract ciprofloxacin, 8-methoxypsoralen [87], and cephalexin [88] (Table 6.2). The drugs were collected after electroporation treatment, with the extracted levels correlating well with those obtained via microdialysis; however, transdermal drug recovery was

TABLE 6.2
Dermatopharmacokinetic Parameters (Mean ± SD, *n* = 6) of Acyclovir and Cephalexin in Hairless Rats as Determined by Microdialysis and by Electroporation and Transcutaneous Sampling and Following IV Administration of the Drugs(for detailed explanations see Refs. [86,88])

	Acyclovir		Cephalexin	
Parameter	Microdialysis	ETS	Microdialysis	ETS
C_{max} (µg/mL)	11.8 ± 2.6	12.0 ± 2.8	12.6 ± 1.9	13.1 ± 1.9
AUC (µg min/mL)	2263.1 ± 326.1	2257.9 ± 254.4	2473.7 ± 202.4	2512.3 ± 250.1
$t_{1/2}$ (min)	78.7 ± 19.3	63.9 ± 15.4	96.3 ± 12.3	106.6 ± 17.8

Sources: Murthy, S.N. and Zhang, S., *J. Dermatol. Sci.*, 49, 249, 2008; Sammeta, S.M. et al., *J. Pharm. Sci.*, 98, 2677, 2009.

drug dependent, being lower for cephalexin and 8-methoxypsoralen. The skin permeability started to recover after a few hours of the permeabilization treatment. The authors propose this technique as less invasive but as efficient as microdialysis to determine dermatopharmacokinetic parameters. In fact, the dermatopharmacokinetic parameters obtained by the two techniques were remarkably similar (Table 6.2). One could expect this method to provide information on the central (plasma) compartment if the distribution and extraction kinetics of the analyte are successfully integrated into the data analysis. It would also be interesting to explore its use as a tool in pharmacokinetic studies carried out on animals, either with preclinical or veterinary purposes. However, it remains to be demonstrated that this technique is less invasive and causes less pain and discomfort than standard blood sampling. Clearly, more is required on the safety and comfort of the pretreatment required on human subjects and research animals. Little has been done on this topic, with exceptions such as the work by Wong et al. [90] who investigated the pain sensation elicited by different electroporation protocols in human volunteers. For example, it was shown that pulsing with a larger electrode (60 pulses, 150 V, 0.1 or 1.0 ms pulse interval), and that increasing the distance between the electrodes caused greater pain. Electrode design was also a key factor; for example, a microelectrode array resulted in almost no perceptible sensation under conditions during which a solid surface electrode caused intolerable pain.

6.4 SONOPHORESIS-MEDIATED SAMPLING

The term sonophoresis is used to describe the enhancement of molecular transport across the skin by means of ultrasound application, low-frequency ultrasound (5–100 kHz) being considered as the most efficient modality of sonophoresis [91,92]. The actual consensus is that transport is enhanced mainly through cavitation in the coupling medium; bubble-induced shock waves and microjet impacts disrupt the SC lipid bilayers. Thermal effects and lipid extraction from the SC during low-frequency sonophoresis also contribute to enhanced transport [91,92]. For example, Merino et al. [93] showed that thermal effects contribute approximately to a quarter of the increase in mannitol flux caused by sonophoresis. Further, it was estimated that 30% of the SC lipids of porcine skin were extracted after a 2 h sonophoretic application (20 kHz, 15 W/cm^2, 0.1:0.9 on/off duty cycle), which also resulted in areas of localized permeabilization visualized by confocal microscopy [94].

However, it should be noticed that application times have been much shorter in human studies. For example, Kost et al. [95] used low-frequency ultrasound to enhance glucose, urea, calcium, and theophylline skin permeability so they could be extracted transdermally in vitro and in vivo. The experimental protocol in human volunteers started with a 1 h hydration of the skin site, followed by a 2 min ultrasound application (20 kHz, 5 s on/5 s off, 10 W/cm^2) using 1% sodium lauryl sulphate (SLS) in saline. Once the skin was permeabilized, the solution was substituted by saline, and vacuum (10 in. Hg) was applied to the collector chamber. Five minute vacuum applications were repeated twice every 30 min and at the end of each the liquid was removed and analyzed for glucose. A calibration factor was obtained using the blood glucose concentration and the extraction flux taking into account a lag time. While this factor varied about 10-fold between patients, the intrasubject variability was less than 20% for the 4 h experiment. Little effects were observed on the skin of the volunteers who reported no pain. These experiments demonstrated that one ultrasound permeabilization short step was sufficient to allow glucose sampling for at least 4 h in diabetic patients.

Cantrell et al. [96] extracted tritiated water from the interstitial fluid (ISF) in hairless rats as a function of the intensity (0–4 W/cm^2), pulse time (0–60 s), and total exposure time (up to 40 min). The nature of cavitation, transient or stable, present during ultrasound application was established by use of a pinducer that measured the 10 kHz subharmonic of the primary ultrasound frequency. Transient cavitation, present in the experiments performed at a higher intensity (3.16 W/cm^2), resulted in increased 3H$_2$O extraction compared with passive controls. This was in agreement with the optical microscopy and SEM examination of the skin after the experiments, which pointed to the SC being disrupted by the highest intensity experiments. However, Mitragotri et al. [97] showed that much shorter applications of ultrasound followed by either passive, vacuum (10 in. Hg), or further low-intensity US could be used to sample analytes from the ISF in rats. A 1 h hydration step was followed by a US pretreatment (20 kHz, 5 s on: 5 s off), which used a 1% solution of SLS in PBS applied until the skin conductivity was raised to (0.6 kΩ · cm^2)$^{-1}$, which typically required 0.25 min. Once the skin was permeabilized, the extraction step took place, the vacuum application being the most efficient procedure. The method extracted ~15 µL of ISF in 15 min and a wide range of analytes (glucose, calcium, urea, triglyceride, lactate, and dextran) were successfully extracted from ISF in a concentration-dependent manner but required the contribution from erroneous sources to be accounted for. A similar setup (US pretreatment and vacuum extraction) was used to follow changes in glucose concentration induced by insulin administration to rats [98]. The results suggested that glucose monitoring could take place but would require individual calibration.

Cook [99] investigated the transdermal extraction of testosterone, 17 β-estradiol, insulin, cortisol, and glucose in sheep and in humans during rest and exercise conditions. Both, a commercial ultrasound and a 10% ethanol gel containing SLS, were tested as vehicles for the permeabilization step that lasted for 1 min of US (20 kHz, 10 W/cm^2, pulsed 5 s on/5 s off) application. A 10 min delay was required before the extraction fluxes were fully established and indicative of blood levels. However, US application was necessary at least every 30 min in order to keep the extraction process working. The extraction of testosterone, 17 β-estradiol, insulin, and cortisol could be further enhanced by application of a 9 V electric field. An interesting contribution of this work was the development of immunosensors for the online measurement of the extracted hormones.

The SonoPrep® device (Sontra Medical) was tested in 10 diabetic patients [100]. The skin was permeabilized in less than 11 s by US application (55 kHz, 12 W) via a 1% SLS in PBS buffer; the degree of permeabilization was monitored by measuring the reduction in skin impedance (Figure 6.6). Three 20 min preliminary glucose sampling steps were done after which a glucose biosensor was applied to the site. The biosensor was coupled to the skin via a hydrogel containing an osmotic extraction buffer and glucose was sampled over 8 h. The patients tolerated well the sonication procedure,

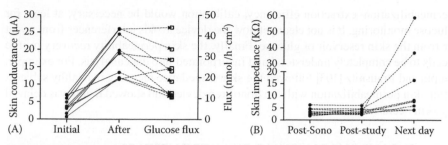

FIGURE 6.6 Effects of ultrasonic pretreatment on the skin conductance, skin impedance, and resultant glucose flux in 10 diabetic patients (100). Two sites per patient were treated with the SonoPrep ultrasonic skin permeation device. (A) The initial skin conductance, mean 4.12 μA was elevated to 18.27 μA in an average sonication time of 6.78 s (range: 3.99–10.49 s). Filled circles joined by a solid line represent the skin conductance, average of two sites per patient, before and after permeabilization. The arrow links the conductance data with the glucose flux data (empty squares) measured for the corresponding patient. (B) The skin impedance (solid circles) was measured at the end of the permeabilization step (PostSono), at the end of the experiment (post-study) and on the next day. Dotted lines link individual data. (Data redrawn from Chuang, H. et al., *Diabetes Technol. Ther.*, 6, 21, 2004.)

which resulted in "no sensation" or "mild sensation"; the treated skin showed a barely perceptible to slight erythema. Glucose was successfully extracted, but the correlation between blood glucose versus sensor reading was poor for three patients.

Lee et al. [101] also performed glucose measurements via electrochemical sensors placed on US-permeabilized skin in vivo in rats. The signal generator was driven at 20 kHz (20% duty cycle, I_{SPTP} = 100 mW/cm² cycle) for 20 min for glucose extraction. The effects of intensity (100–300 mW/cm²) and duty cycle (15% or 20%) on the skin were investigated. The experiments performed with 200 mW/cm²–20% and 300 mW/cm²–20% conditions produced gross histological lesions (large spaces or cavitated areas, disruption of sebaceous glands and hair follicles, and degeneration of collagen) localized under the transducers and more severe for the 300 mW/cm²–20% group.

Echo Therapeutics (previously Sontra Medical) announced in July 2008 positive results for their clinical studies using the device Symphony™ tCGM (transdermal continuous glucose monitoring) in type I and II diabetic patients [102]. This system incorporates wireless transmission and transdermal biosensor technologies as well as the Prelude™ SkinPrep System that provides skin permeation control and has been cleared by the FDA.

In summary, low-frequency US allows a very fast permeabilization of the skin, in less than a minute, after which extraction fluxes are typically higher than those achieved with iontophoresis. Further, it expands the range of analytes that can be sampled transdermally to higher molecular weight substances such as hormones. However, the same US pretreatment does not provide equivalent permeabilization in different individuals, so a normalization method is typically required, for example, by measurement of skin impedance or conductance (Figure 6.6). Once the skin is permeabilized, an additional process (osmotic buffer, vacuum application) is necessary to perform the extraction step. Because of the variability observed in the permeabilization–extraction efficiency, calibration would be necessary, at least for glucose monitoring. It is not clear yet how the device avoids interferences from sweat or from the skin reservoir of glucose. Finally, the skin permeability recovery profile needs to be completely understood and incorporated into data analysis. For example, Gupta and Prausnitz [103] followed the skin impedance recovery in healthy subjects after skin permeabilization with the SonoPrep device and showed that it was delayed by skin occlusion.

6.5 CUTANEOUS SAMPLING BY MICRONEEDLES

Microneedles, available in different formats, materials, and structures, aim to bypass the SC or main barrier to skin transport without causing the pain typically associated with standard hypodermic needles [104–106]. Kaushik et al. [107] compared the pain experienced by volunteers during insertion of a 26 gauge hypodermic needle, a 3×3 mm array of 400–150 µm height microneedles, and a negative control (a silicon wafer). The sensation caused by the microneedles and the negative control was indistinguishable and significantly less than that caused by the hypodermic needle. Ideally, microneedles should pierce the skin reliably without breaking or being clogged, and reach no deeper than the epidermis, allowing transport of fluid but not of harmful agents [108,109]. Other significant properties are their potential toxicity, irritation, and recovery rates. For example, some work has looked into the effect of length and shape of microneedles on TEWL and irritation [110]. Other studies have investigated pain intensity, sensory perception, and TEWL after use of various applicators, microneedles of different lengthes and compared them with the use of a hypodermic 25-G needle [111]. An electrical applicator has been proposed so the microneedles penetrate the skin with a predefined velocity, which results in improved piercing and drug delivery compared to manual application [112].

Arrays of hollow microneedles with different design ("volcano", "micro-hypodermic," and "snake fang") were fabricated and tested by Mukerjee et al. [113]. The 20×20 arrays of microneedles were effective in puncturing the skin and sampling ISF via capillary action. This was possible but required 20–30 min for the needle and channels to be filled initially. Interestingly, the performance of the microneedles was conditioned by their design. For example, the "snake fang" skin successfully extracted ISF from the skin while the "micro-hypodermic" design was prone to blockage during application. According to the authors, the pressure needed to pierce the skin was similar to that required to apply a Band-Aid (~1.5 N ± 0.25).

A different approach has looked into glass microneedles whose function was simply to puncture the skin; the hole(s) created were subsequently used to sample ISF via vacuum application. The procedure was used in rats and healthy volunteers and could successfully follow glucose blood concentration [114]. A single microneedle was inserted 6–10 times within a 1 cm^2 area after which vacuum (200–500 mm Hg) was applied for 2–10 min. This protocol extracted 1–10 µL of ISF, which was collected and analyzed with a standard glucose monitoring device. Some erythema was observed at the end of the experiments. A key advantage of this process is that ISF is directly harvested so glucose, or any other analyte of interest, is directly sampled, that is, avoiding dilution steps into a collector chamber and facilitating quantification.

Evaporation of ISF at the skin surface caused some data distortion but was avoided by placing a film of vacuum oil onto the skin. It is not yet clear whether this extraction procedure would require calibration. Subsequent work [115] looked into the use of standard glucose blood measuring devices for ISF samples. The results indicated that measuring devices would need to be specifically adapted for ISF analysis, as use of standard glucose measuring devices for ISF samples resulted in biased measurements.

Other studies have looked into microporation of the skin by different means. For example, micropores can be created by thermal ablation of the skin via a laser beam (SpectRx, Inc.) followed by application of a pressure gradient to harvest ISF [116]. Diabetic patients were administered sodium fluorescein via IV bolus and ISF was subsequently harvested at two different sites. The distribution of fluorescein into ISF was quickly tracked from the ISF extracts. This study clearly shows that transdermal sampling takes place in a different compartment than plasma, and that a good understanding of the distribution kinetics is required to interpret extraction data. The same device was also tested in 110 children (2–18 years old) [117] who reported a similar sensation for the microporation and the standard fingerstick procedures. Nonetheless, the authors claim the new system would be performed only once for a 1–3 day period during which multiple samples could be collected. However, it is not clear how often the calibration procedure (daily in the study) would be required in practice and glucose tracking could be optimized. In subsequent work [118], the SpectRx system was used to sample insulin-like growth factor-I (IGF-I) in transdermal fluid. Total IGF-I was lower (~25%) in transdermal fluid than in serum, a finding in agreement with previous measurements done via the blister technique. Thus, sampling ISF or transdermal interstitial fluid may provide information relevant to drug concentration in the tissue compartment, a value that for some drugs can be more relevant to therapeutic and toxic effects than blood levels.

An alternative procedure [119] used micro-heater elements to ablate the SC and subsequently collect ISF; the analytes were harvested into a physiological sampling fluid released from a reservoir and then transferred to the glucose measuring membrane. Preliminary tests on surrogate human graft skin (Apigraf, Organogenesis) showed that a 50 mJ pulse was sufficient to create a 50 μm diameter hole; the ablation temperature was 130°C and SC vaporization occurred in less than 33 ms, with ISF arriving to the skin surface in ~99 ms. In another study [120], blood could be sampled in less than 20 s via ~200 μm deep microconduits caused by microscission, which caused the patients less pain than the standard glucose lancet.

6.6 CONCLUDING REMARKS

Transdermal sampling constitutes an interesting and advantageous alternative for noninvasive sampling. Passive diffusion is inefficient for most molecules, requiring lengthy sampling periods and providing primarily qualitative information. Enhancement techniques have been investigated such as iontophoresis, electroporation, ultrasound and microneedles with only one device, the GlucoWatch Biographer, having reached the market. A thorough comparison of the available techniques should consider at least (a) what is known about the mechanism of extraction (efficiency, transport control, factors determining extraction fluxes, variability, reproducibility, range of application); (b) safety of the device after single and multiple applications, discomfort, pain, irritation, need for sterilization, risk of infection, and length of the healing process; (c) need for calibration and feasible calibration procedures; (d) contribution from sources (sweat and skin reservoir) other than ISF to the extracted amounts; (e) number of additional processes required, for example, normalization of the permeabilization step, ISF extraction by capillary forces or vacuum application, and the need for built-in sensors; (f) kinetics of extraction, for example, length of the stabilization time for iontophoresis, recovery of skin permeability after electroporation and ultrasound pretreatment, healing process in microporated skin; and (g) cost-effectiveness and financial viability. In addition, a good understanding of the pharmacokinetic and metabolic pathways concerning the analyte of interest and of the clinical considerations and aims (assessment of exposure, therapeutic monitoring, pharmacokinetic profiling, and continuous monitoring) are crucial to decide on the requirements that a noninvasive sampling technique must meet for a chosen compound.

The advances observed in transdermal sampling since the 1980s are considerable and some impressive technology such as the GlucoWatch Biographer has been developed. Wide-ranging work in the field has provided a much better understanding of the complexities and challenges posed by truly noninvasive transdermal sampling for different applications and analytes. Current expertise can guide the development of future improved devices and technologies, which at the very end must prove their superiority with respect to the humble hypodermic needle.

REFERENCES

1. Peck, C. C., K. Lee, and C. E. Becker. 1981. Continuous transepidermal drug collection: Basis for use in assessing drug intake and pharmacokinetics. *J Pharmacokinet Biopharm* 9:41–58.
2. Peck, C. C., B. J. Bolden, R. G. Alimirez et al. 1987. Outward transdermal migration of theophylline. *Pharmacol Skin* 1:201–208.
3. Peck, C. C., D. P. Conner, B. J. Bolden et al. 1988. Outward transcutaneous chemical migration: Implications for diagnostics and dosimetry. *Skin Pharmacol* 1:14–23.
4. Conner, D. P., R. G. Almirez, P. Rhyne, K. Zamani, B. J. Bolden, and C. C. Peck. 1989. Transcutaneous collection of theophylline: Constancy and linearity of skin permeability. *Skin Pharmacol* 2:155–161.
5. Bradley, C. R., R. G. Almirez, D. P. Conner, P. R. Rhyne, and C. C. Peck. 1990. Noninvasive transdermal chemical collection. II. In vitro and in vivo skin permeability studies. *Skin Pharmacol* 3:227–235.
6. Murphy, M. G., C. C. Peck, D. P. Conner, K. Zamani, G. B. Merenstein, and D. Rodden. 1990. Transcutaneous theophylline collection in preterm infants. *Clin Pharmacol Ther* 47:427–434.
7. Conner, D. P., E. Millora, K. Zamani et al. 1991. Transcutaneous chemical collection of caffeine in normal subjects: Relationship to area under the plasma concentration-time curve and sweat production. *J Invest Dermatol* 96:186–190.
8. Pellett, M. A., J. Hadgraft, and M. S. Roberts. 1999. The back diffusion of glucose across human skin in vitro. *Int J Pharm* 193:27–35.
9. Kintz, P., A. Tracqui, and P. Mangin. 1996. Sweat testing for benzodiazepines. *J Forensic Sci* 41:851–854.

10. Cone, E. J., M. J. Hillsgrove, A. J. Jenkins, R. M. Keenan, and W. D. Darwin. 1994. Sweat testing for heroin, cocaine, and metabolites. *J Anal Toxicol* 18:298–305.

11. Kintz, P., R. Brenneisen, P. Bundeli, and P. Mangin. 1997. Sweat testing for heroin and metabolites in a heroin maintenance program. *Clin Chem* 43:736–739.

12. Kidwell, D. A., J. D. Kidwell, F. Shinohara et al. 2003. Comparison of daily urine, sweat, and skin swabs among cocaine users. *Forensic Sci Int* 133:63–78.

13. Kintz, P., A. Tracqui, C. Jamey, and P. Mangin. 1996. Detection of codeine and phenobarbital in sweat collected with a sweat patch. *J Anal Toxicol* 20:197–201.

14. Kintz, P., A. Tracqui, P. Mangin, and Y. Edel. 1996. Sweat testing in opioid users with a sweat patch. *J Anal Toxicol* 20:393–397.

15. Burns, M. and R. C. Baselt. 1995. Monitoring drug use with a sweat patch: An experiment with cocaine. *J Anal Toxicol* 19:41–48.

16. Kalia, Y. N., A. Naik, J. Garrison, and R. H. Guy. 2004. Iontophoretic drug delivery. *Adv Drug Deliv Rev* 56:619–658.

17. Delgado-Charro, M. B. and R. H. Guy. 2001. Transdermal iontophoresis for controlled drug delivery and non-invasive monitoring. *Stp Pharma Sci* 11:403–414.

18. Delgado-Charro, M. B. 2009. Recent advances on transdermal iontophoretic drug delivery and non-invasive sampling. *J Drug Deliv Sci Tech* 19:75–88.

19. Sage, B. H. and J. E. Riviere. 1992. Model systems in iontophoresis–Transport efficacy. *Adv. Drug Deliv Rev* 9:265–287.

20. Sieg, A. and V. Wascotte. 2009. Diagnostic and therapeutic applications of iontophoresis. *J Drug Target*. Doi: 10.101080/10611860903089750.

21. Leboulanger, B., R. H. Guy, and M. B. Delgado-Charro. 2004. Reverse iontophoresis for non-invasive transdermal monitoring. *Physiol Meas* 25:R35–R50.

22. Ledger, P. W. 1992. Skin biological issues in electrically enhanced transdermal delivery. *Adv. Drug Deliv Rev* 9:289–307.

23. Mudry, B., P. A. Carrupt, R. H. Guy, and M. B. Delgado-Charro. 2007. Quantitative structure-permeation relationship for iontophoretic transport across the skin. *J Control Release* 122:165–172.

24. Mudry, B., R. H. Guy, and M. Begoña Delgado-Charro. 2006. Prediction of iontophoretic transport across the skin. *J Control Release* 111:362–367.

25. Mudry, B., R. H. Guy, and M. B. Delgado-Charro. 2006. Transport numbers in transdermal iontophoresis. *Biophys J* 90:2822–2830.

26. Mudry, B., R. H. Guy, and M. B. Delgado-Charro. 2006. Electromigration of ions across the skin: Determination and prediction of transport numbers. *J Pharm Sci* 95:561–569.

27. Phipps, J. B., R. V. Padmanabhan, and G. A. Lattin. 1989. Iontophoretic delivery of model inorganic and drug ions. *J Pharm Sci* 78:365–369.

28. Leboulanger, B., J. M. Aubry, G. Bondolfi, R. H. Guy, and M. B. Delgado-Charro. 2004. Lithium monitoring by reverse iontophoresis in vivo. *Clin Chem* 50:2091–2100.

29. Sieg, A., R. H. Guy, and M. B. Delgado-Charro. 2004. Noninvasive glucose monitoring by reverse iontophoresis in vivo: Application of the internal standard concept. *Clin Chem* 50:1383–1390.

30. Nixon, S., A. Sieg, M. B. Delgado-Charro, and R. H. Guy. 2007. Reverse iontophoresis of L-lactate: In vitro and in vivo studies. *J Pharm Sci* 96:3457–3465.

31. Leboulanger, B., R. H. Guy, and M. B. Delgado-Charro. 2004. Non-invasive monitoring of phenytoin by reverse iontophoresis. *Eur J Pharm Sci* 22:427–433.

32. Santi, P. and R. H. Guy. 1996. Reverse iontophoresis—Parameters determining electroosmotic flow: I. pH and ionic strength. *J Control Release* 38:159–165.

33. Santi, P. and R. H. Guy. 1996. Reverse iontophoresis—Parameters determining electroosmotic flow. II: Electrode chamber formulation. *J Control Release* 42:29–36.

34. Merino, V., A. Lopez, D. Hochstrasser, and R. H. Guy. 1999. Noninvasive sampling of phenylalanine by reverse iontophoresis. *J Control Release* 61:65–69.

35. Glikfeld, P., R. S. Hinz, and R. H. Guy. 1989. Noninvasive sampling of biological fluids by iontophoresis. *Pharm Res* 6:988–990.
36. Rao, G., P. Glikfeld, and R. H. Guy. 1993. Reverse iontophoresis: Development of a noninvasive approach for glucose monitoring. *Pharm Res* 10:1751–1755.
37. Rao, G., R. H. Guy, P. Glikfeld et al. 1995. Reverse iontophoresis: Noninvasive glucose monitoring in vivo in humans. *Pharm Res* 12:1869–1873.
38. Green, P. G., R. S. Hinz, A. Kim, F. C. Szoka, and R. H. Guy. 1991. Iontophoretic delivery of a series of tripeptides across the skin in vitro. *Pharm Res* 8:1121–1127.
39. Tamada, J. A., N. J. Bohannon, and R. O. Potts. 1995. Measurement of glucose in diabetic subjects using noninvasive transdermal extraction. *Nat Med* 1:1198–1201.
40. Potts, R. O., J. A. Tamada, and M. J. Tierney. 2002. Glucose monitoring by reverse iontophoresis. *Diabetes Metab Res Rev* 18 Suppl 1:S49–S53.
41. Tierney, M. J., J. A. Tamada, R. O. Potts, L. Jovanovic, and S. Garg. 2001. Clinical evaluation of the GlucoWatch biographer: A continual, non-invasive glucose monitor for patients with diabetes. *Biosens Bioelectron* 16:621–629.
42. Tierney, M. J., J. A. Tamada, R. O. Potts et al. 2000. The GlucoWatch biographer: A frequent automatic and noninvasive glucose monitor. *Ann Med* 32:632–641.
43. Pitzer, K. R., S. Desai, T. Dunn, S. et al. 2001. Detection of hypoglycemia with the GlucoWatch biographer. *Diabetes Care* 24:881–885.
44. Garg, S. K., R. O. Potts, N. R. Ackerman, S. J. Fermi, J. A. Tamada, and H. P. Chase. 1999. Correlation of fingerstick blood glucose measurements with GlucoWatch biographer glucose results in young subjects with type 1 diabetes. *Diabetes Care* 22:1708–1714.
45. Tamada, J. A., N. J. Bohannon, and R. O. Potts. 1995. Measurement of glucose in diabetic subjects using noninvasive transdermal extraction. *Nat Med* 1:1198–1201.
46. Sieg, A., R. H. Guy, and M. B. Delgado-Charro. 2004. Simultaneous extraction of urea and glucose by reverse iontophoresis in vivo. *Pharm Res* 21:1805–1810.
47. Hagstrom-Toft, E., S. Enoksson, E. Moberg, J. Bolinder, and P. Arner. 1997. Absolute concentrations of glycerol and lactate in human skeletal muscle, adipose tissue, and blood. *Am J Physiol* 273:E584–E592.
48. Hagstrom, E., P. Arner, U. Ungerstedt, and J. Bolinder. 1990. Subcutaneous adipose tissue: A source of lactate production after glucose ingestion in humans. *Am J Physiol* 258:E888–E893.
49. Wascotte, V., M. B. Delgado-Charro, E. Rozet et al. 2007. Monitoring of urea and potassium by reverse iontophoresis in vitro. *Pharm Res* 24:1131–1137.
50. Wascotte, V., M. B. Delgado-Charro, R. H. Guy, and V. Preat. 2004. Monitoring renal function by reverse iontophoresis. *Eur J Pharm Sci* 23:S57–S57.
51. Wascotte, V., P. Caspers, J. de Sterke, M. Jadoul, R. H. Guy, and V. Preat. 2007. Assessment of the "Skin reservoir" of urea by confocal raman microspectroscopy and reverse iontophoresis in vivo. *Pharm Res* 24:1897–1901.
52. Sieg, A., F. Jeanneret, M. Fathi, D. Hochstrasser, S. Rudaz, J. L. Veuthey, R. H. Guy, and M. B. Delgado-Charro. 2009. Extraction of amino acids by reverse iontophoresis in vivo. *Eur J Pharm Biopharm*. 72, 226–231.
53. Sieg, A., F. Jeanneret, M. Fathi, D. Hochstrasser, S. Rudaz, J. L. Veuthey, R. H. Guy, and M. B. Delgado-Charro. 2008. Extraction of amino acids by reverse iontophoresis: Simulation of therapeutic monitoring in vitro. *Eur J Pharm Biopharm* 70:908–913.
54. Bouissou, C. C., J. P. Sylvestre, R. H. Guy, M. B. Delgado-Charro. 2009. Reverse iontophoresis of amino acids: Identification and separation of stratum corneum and subdermal sources in vitro. *Pharm Res*. 26(12), 2630–2638.
55. The Diabetes Control and Complications Trial Research Group. 1993. The effect of intensive treatment of diabetes on the development and progression of long-term complications in insulin-dependent diabetes mellitus. *N Engl J Med* 329:977–986.

56. Potts, R. O., J. A. Tamada, and M. J. Tierney. 2002. Glucose monitoring by reverse iontophoresis. *Diabetes Metab Res Rev* 18:S49–S53.
57. Tamada, J. A., S. Garg, L. Jovanovic, K. R. Pitzer, S. Fermi, and R. O. Potts. 1999. Noninvasive glucose monitoring: Comprehensive clinical results. Cygnus Research Team. *JAMA* 282:1839–1844.
58. Eastman, R. C., H. P. Chase, B. Buckingham et al. 2002. Use of the GlucoWatch biographer in children and adolescents with diabetes. *Pediatr Diabetes* 3:127–134.
59. Eastman, R. C., A. D. Leptien, and H. P. Chase. 2003. Cost-effectiveness of use of the GlucoWatch Biographer in children and adolescents with type 1 diabetes: A preliminary analysis based on a randomized controlled trial. *Pediatr Diabetes* 4:82–86.
60. Tura, A., A. Maran, and G. Pacini. 2007. Non-invasive glucose monitoring: Assessment of technologies and devices according to quantitative criteria. *Diabetes Res Clin Pract* 77:16–40.
61. Tierney, M. J., Y. Jayalakshmi, N. A. Parris, M. P. Reidy, C. Uhegbu, and P. Vijayakumar. 1999. Design of a biosensor for continual, transdermal glucose monitoring. *Clin Chem* 45:1681–1683.
62. The diabetes research in children network (DirecNet) study group. 2004. Accuracy of the GlucoWatch G2 Biographer and the continuous glucose monitoring system during hypoglycemia: Experience of the Diabetes Research in Children Network. *Diabetes Care* 27:722–726.
63. Nunnold, T., S. R. Colberg, M. T. Herriott, and C. T. Somma. 2004. Use of the noninvasive GlucoWatch Biographer during exercise of varying intensity. *Diabetes Technol Ther* 6:454–462.
64. Tamada, J. A., T. L. Davis, A. D. Leptien et al. 2004. The effect of preapplication of corticosteroids on skin irritation and performance of the GlucoWatch G2 Biographer. *Diabetes Technol Ther* 6:357–367.
65. Mendosa, D., 2006, The Glucowatch Biographer—Update, http://www.diabetesnet.com/diabetes_technology/glucowatch.php
66. Mize, N. K., M. Buttery, P. Daddona, C. Morales, and M. Cormier. 1997. Reverse iontophoresis: Monitoring prostaglandin E2 associated with cutaneous inflammation in vivo. *Exp Dermatol* 6:298–302.
67. Gruemer, H. D., G. F. Grannis, L. B. Hetland, and M. L. Costantini. 1971. Amino acid transport and mental retardation. *Clin Chem* 17:1129–1131.
68. Longo, N., S. K. Li, G. Yan et al. 2007. Noninvasive measurement of phenylalanine by iontophoretic extraction in patients with phenylketonuria. *J Inherit Metab Dis* 30:910–915.
69. Sylvestre, J. P. 2007. Applications of iontophoresis in sports medicine. *PhD Thesis.* University of Bath.
70. Degim, I. T., S. Ilbasmis, R. Dundaroz, and Y. Oguz. 2003. Reverse iontophoresis: A noninvasive technique for measuring blood urea level. *Pediatr Nephrol* 18:1032–1037.
71. Wascotte, V., E. Rozet, A. Salvaterra et al. 2008. Non-invasive diagnosis and monitoring of chronic kidney disease by reverse iontophoresis of urea in vivo. *Eur J Pharm Biopharm* 69:1077–1082.
72. Djabri, A., R. H. Guy, and M. B. Delgado-Charro. 2009. Transdermal reverse iontophoresis of iohexol as a tool for minimally-invasive monitoring of renal function. *10th International Congress of Therapeutic Drug Monitoring & Clinical Toxicology, Nice, France, 9–14th September 2007.*
73. Djabri, A., W. Van't Hoff, P. Brock, I. C. K. Wong, R. H. Guy, and M. B. Delgado-Charro. Non-invasive assessment of renal function via transdermal reverse iontophoresis of iohexol: A pilot study. *11th International Congress of TDM & Clinical Toxicology, Montreal, Canada, 3–8th October 2009.*

74. Djabri, A. 2009. Iontophoresis in paediatric medicine: Non-invasive drug delivery and monitoring applications. *PhD Thesis*. University of Bath.
75. Billat, V. L., P. Sirvent, G. Py, J. P. Koralsztein, and J. Mercier. 2003. The concept of maximal lactate steady state: A bridge between biochemistry, physiology and sport science. *Sports Med* 33:407–426.
76. Kellum, J. A. 1998. Lactate and pHi: Our continued search for markers of tissue distress. *Crit Care Med* 26:1783–1784.
77. Delgado-Charro, M. B. and R. H. Guy. 2003. Transdermal reverse iontophoresis of valproate: A noninvasive method for therapeutic drug monitoring. *Pharm Res* 20:1508–1513.
78. Wascotte, V., B. Leboulanger, R. H. Guy, and M. B. Delgado-Charro. 2005. Reverse iontophoresis of lithium: Electrode formulation using a thermoreversible polymer. *Eur J Pharm Biopharm* 59:237–240.
79. Leboulanger, B., M. Fathi, R. H. Guy, and M. B. Delgado-Charro. 2004. Reverse iontophoresis as a noninvasive tool for lithium monitoring and pharmacokinetic profiling. *Pharm Res* 21:1214–1222.
80. Nicoli, S. and P. Santi. 2006. Transdermal delivery of aminoglycosides: Amikacin transport and iontophoretic non-invasive monitoring. *J Control Release* 111:89–94.
81. Sekkat, N., A. Naik, Y. N. Kalia, P. Glikfeld, and R. H. Guy. 2002. Reverse iontophoretic monitoring in premature neonates: Feasibility and potential. *J Control Release* 81:83–89.
82. M. J. Tierney, H. L. Kim, M. D. Burns, J. A. Tamada, R. O. Potts. 2000. Electroanalysis of glucose in transcutaneously extracted samples. *Electroanalysis* 12 666–671.
83. Sieg, A., R. H. Guy, and M. B. Delgado-Charro. 2003. Reverse iontophoresis for noninvasive glucose monitoring: the internal standard concept. *J Pharm Sci* 92:2295–2302.
84. Escobar-Chavez, J. J. D., D. Bonilla-Marti Nez, M. A. Villegas-Gonzalez, and A. L. Revilla-Vazquez. 2009. Electroporation as an efficient physical enhancer for skin drug delivery. *J Clin Pharmacol*. Doi:10.1177/0091270009344984.
85. Murthy, S. N., Y. L. Zhao, S. W. Hui, and A. Sen. 2005. Electroporation and transcutaneous extraction (ETE) for pharmacokinetic studies of drugs. *J Control Release* 105:132–141.
86. Murthy, S. N. and S. Zhang. 2008. Electroporation and transcutaneous sampling (ETS) of acyclovir. *J Dermatol Sci* 49:249–251.
87. Sammeta, S. M., S. R. Vaka, and S. N. Murthy. 2009. Transcutaneous sampling of ciprofloxacin and 8-methoxypsoralen by electroporation (ETS technique). *Int J Pharm* 369:24–29.
88. Sammeta, S. M., S. R. Vaka, and S. N. Murthy. 2009. Dermal drug levels of antibiotic (cephalexin) determined by electroporation and transcutaneous sampling (ETS) technique. *J Pharm Sci* 98:2677–2685.
89. Srinivasa Murthy, S., V. Siva Ram Kiran, S. Mathur, and S. Narasimha Murthy. 2008. Noninvasive transcutaneous sampling of glucose by electroporation. *J Diabetes Sci Technol* 2:250–254.
90. Wong, T. W., C. H. Chen, C. C. Huang, C. D. Lin, and S. W. Hui. 2006. Painless electroporation with a new needle-free microelectrode array to enhance transdermal drug delivery. *J Control Release* 110:557–565.
91. Mitragotri, S. and J. Kost. 2004. Low-frequency sonophoresis: A review. *Adv Drug Deliv Rev* 56:589–601.
92. Merino, G., Y. N. Kalia, and R. H. Guy. 2003. Ultrasound-enhanced transdermal transport. *J Pharm Sci* 92:1125–1137.
93. Merino, G., Y. N. Kalia, M. B. Delgado-Charro, R. O. Potts, and R. H. Guy. 2003. Frequency and thermal effects on the enhancement of transdermal transport by sonophoresis. *J Control Release* 88:85–94.

94. Alvarez-Roman, R., G. Merino, Y. N. Kalia, A. Naik, and R. H. Guy. 2003. Skin permeability enhancement by low frequency sonophoresis: Lipid extraction and transport pathways. *J Pharm Sci* 92:1138–1146.
95. Kost, J., S. Mitragotri, R. A. Gabbay, M. Pishko, and R. Langer. 2000. Transdermal monitoring of glucose and other analytes using ultrasound. *Nat Med* 6:347–350.
96. Cantrell, J. T., M. J. McArthur, and M. V. Pishko. 2000. Transdermal extraction of interstitial fluid by low-frequency ultrasound quantified with $3H_2O$ as a tracer molecule. *J Pharm Sci* 89:1170–1179.
97. Mitragotri, S., M. Coleman, J. Kost, and R. Langer. 2000. Transdermal extraction of analytes using low-frequency ultrasound. *Pharm Res* 17:466–470.
98. Mitragotri, S., M. Coleman, J. Kost, and R. Langer. 2000. Analysis of ultrasonically extracted interstitial fluid as a predictor of blood glucose levels. *J Appl Physiol* 89:961–966.
99. Cook, C. J. 2002. Rapid noninvasive measurement of hormones in transdermal exudate and saliva. *Physiol Behav* 75:169–181.
100. Chuang, H., E. Taylor, and T. W. Davison. 2004. Clinical evaluation of a continuous minimally invasive glucose flux sensor placed over ultrasonically permeated skin. *Diabetes Technol Ther* 6:21–30.
101. Lee, S., V. Nayak, J. Dodds, M. Pishko, and N. B. Smith. 2005. Glucose measurements with sensors and ultrasound. *Ultrasound Med Biol* 31:971–977.
102. Echo therapeutics. 2008. Echo therapeutics announces positive results of a clinical study. http://www.echotx.com/files/attachments/echotx/Newsroom/Echo_Abstract_Presentation_MR1_Study_Final%20(2).pdf
103. Gupta, J. and M. R. Prausnitz. 2009. Recovery of skin barrier properties after sonication in human subjects. *Ultrasound Med Biol* 35:1405–1408.
104. Prausnitz, M. R. 2004. Microneedles for transdermal drug delivery. *Adv Drug Deliv Rev* 56:581–587.
105. McAllister, D. V., P. M. Wang, S. P. Davis et al. 2003. Microfabricated needles for transdermal delivery of macromolecules and nanoparticles: Fabrication methods and transport studies. *Proc Natl Acad Sci USA* 100:13755–13760.
106. Sivamani, R. K., D. Liepmann, and H. I. Maibach. 2007. Microneedles and transdermal applications. *Expert Opin Drug Deliv* 4:19–25.
107. Kaushik, S., A. H. Hord, D. D. Denson et al. 2001. Lack of pain associated with microfabricated microneedles. *Anesth Analg* 92:502–504.
108. Zahn, J. D., Y. C. Hsieh, and M. Yang. 2005. Components of an integrated microfluidic device for continuous glucose monitoring with responsive insulin delivery. *Diabetes Technol Ther* 7:536–545.
109. Friedl, K. E. 2005. Analysis: Optimizing microneedles for epidermal access. *Diabetes Technol Ther* 7:546–548.
110. Bal, S. M., J. Caussin, S. Pavel, and J. A. Bouwstra. 2008. In vivo assessment of safety of microneedle arrays in human skin. *Eur J Pharm Sci* 35:193–202.
111. Haq, M. I., E. Smith, D. N. John et al. 2009. Clinical administration of microneedles: Skin puncture, pain and sensation. *Biomed Microdevices* 11:35 47.
112. Verbaan, F. J., S. M. Bal, D. J. van den Berg et al. 2008. Improved piercing of microneedle arrays in dermatomed human skin by an impact insertion method. *J Control Release* 128:80–88.
113. Mukerjee, E. V., S. D. Collins, R. R. Isseroff, and R. L. Smith. 2004. Microneedle array for transdermal biological fluid extraction and in situ analysis. *Sens Actuators A Phys* 114:267–275.
114. Wang, P. M., M. Cornwell, and M. R. Prausnitz. 2005. Minimally invasive extraction of dermal interstitial fluid for glucose monitoring using microneedles. *Diabetes Technol Ther* 7:131–141.

115. Vesper, H. W., P. M. Wang, E. Archibold, M. R. Prausnitz, and G. L. Myers. 2006. Assessment of trueness of a glucose monitor using interstitial fluid and whole blood as specimen matrix. *Diabetes Technol Ther* 8:76–80.

116. Smith, A., D. Yang, H. Delcher, J. Eppstein, D. Williams, and S. Wilkes. 1999. Fluorescein kinetics in interstitial fluid harvested from diabetic skin during fluorescein angiography: Implications for glucose monitoring. *Diabetes Technol Ther* 1:21–27.

117. Burdick, J., P. Chase, M. Faupel, B. Schultz, and S. Gebhart. 2005. Real-time glucose sensing using transdermal fluid under continuous vacuum pressure in children with type 1 diabetes. *Diabetes Technol Ther* 7:448–455.

118. Nindl, B. C., A. P. Tuckow, J. A. Alemany et al. 2006. Minimally invasive sampling of transdermal body fluid for the purpose of measuring insulin-like growth factor-I during exercise training. *Diabetes Technol and Ther* 8:244–252.

119. Paranjape, M., J. Garra, S. Brida, T. Schneider, R. White, and J. Currie. 2003. A PDMS dermal patch for non-intrusive transdermal glucose sensing. *Sens Actuators A Phys* 104:195–204.

120. Herndon, T. O., S. Gonzalez, T. R. Gowrishankar, R. R. Anderson, and J. C. Weaver. 2004. Transdermal microconduits by microscission for drug delivery and sample acquisition. *BMC Med* 2:12. http://www.biomedcentral.com/1741-7015.2

7 Spectroscopic Techniques in Dermatokinetic Studies

Georgios N. Stamatas

CONTENTS

7.1 INTRODUCTION TO SPECTROSCOPIC TECHNIQUES USED IN DERMATOKINETIC STUDIES

The science of dermatokinetics is defined as the study of the penetration dynamics of a topically applied active (usually a drug) through the skin over time. To this end, it is obvious that accurate methods are needed to quantify the amount of active in the skin tissue. Since the aim is to study the same sample (skin) over time, it is necessary that the measuring methods are nondestructive. Furthermore, in order not to disturb the naturally evolving kinetics of drug penetration, these methods need not only be nondestructive but also noninvasive. Finally, although in vitro models (such as skin equivalents) or ex vivo models (excised skin from cadavers or from surgeries) are useful as a first step to understand dermatokinetics, the ultimate goal is to be able to study drug penetration in human skin in vivo. The techniques that fulfill all the above requirements are based on the principle of interrogating the skin tissue using an optical probe. In other words, while light* is allowed to travel through and interact with the sample, part of the light is collected and analyzed. This analysis of the collected light, known as spectroscopy, is able to provide us with information regarding the concentration (and in some cases the distribution) of the drug in the tissue at a particular point in time.

It is important to note that non-optic methods have previously been used, including minimally invasive sampling of the skin surface by tape stripping followed by further chemical analysis (for a review, see Escobar-Chavez et al. 2008), for example, high-performance liquid chromatography (HPLC). Since historically these methods have been developed earlier, they are often used for validation of the newly developed spectroscopic techniques. However, one needs to keep in mind that the sampling methods are limited to the information at the surface of the tissue and that potentially the act of sampling is affecting the natural course of drug kinetics since part of the drug compartment at the skin surface is removed.

Although the principles of the methods described in this chapter were originally based on point measurements, where the interrogated tissue volume is optically averaged, recent advancements in hardware have allowed for such techniques to be used in imaging mode. In imaging, information of a grid of adjacent small volumes in the tissue is compiled into two-dimensional maps. In this chapter, we examine some applications of imaging technologies mostly related to Raman or higher-order methods.

We will begin with examining diffuse reflectance spectroscopy (DRS), a technique based on selective absorption of light by the drug in the tissue. Taking advantage of the fact that some drugs fluoresce, we will examine the use of fluorescence spectroscopy (FS). Vibrational spectroscopies include infrared (IR)- and Raman-based methods. They are both based on excitation of vibration modes between the chemical bonds that make up a molecule and can result in characteristic spectra of the molecules of interest. Both of these methods have been shown to be very useful in dermatokinetics and we will review some examples of their application. We will further examine photoacoustic spectroscopy (PAS) along with the development of new methods based on nonlinear (higher-order) optics and their applications in dermatokinetics. Finally, we will close with a review and recommendations for future research.

7.2 FUNDAMENTALS OF TISSUE SPECTROSCOPY

Humans are very visual, that is, we receive most of the information about the world through our eyes. Light and how light interacts with the world comes to us intuitively. Physics and in particular optics has explained in detail the interactions of light with matter. Understanding these interactions enables their utilization in solving problems such as what is the concentration of a particular drug in the skin.

Light can be thought of as energy waves in space and time or as particles (photons) of a specified amount of energy traveling through space with a specified velocity.

---

* For the purposes of this chapter we will use the term "light" in its broad sense of electromagnetic radiation that covers the visible spectral region as well as the regions adjacent to it: ultraviolet (UV) and infrared (IR).

Let us consider initially the second definition. A beam of light particles is directed to the skin surface. Upon incidence, a small part (about 4%) of the particles will be deflected away from the surface and the rest will penetrate it. The light particles can interact with the skin components (as well as with any externally applied drugs that have penetrated the skin) in several ways. They can change the direction of travel, be absorbed and extinguished (transformed to heat), or be absorbed and reemitted with the same or different energy.

7.2.1 LIGHT SCATTERING

A change in the direction of travel as a result of light–matter interaction is called scattering. The efficiency with which a particle is able to scatter light is given by the scattering coefficient and the change in direction is given by the anisotropy parameter. Light scattering in a turbid material, such as skin, is responsible for diffusing the light into the tissue (therefore increasing the volume of tissue that is illuminated) and for changing the direction of travel enough number of times so that part of that light is able to reemerge at the surface of the skin and get collected and analyzed. Hence, light scattering is important for all types of in vivo spectroscopy.

Light scattering is also important because it is a major factor determining how deep light would travel into the tissue. Strong scattering of blue light, for example, limits its penetration to approximately the papillary dermis (top layers of dermis), whereas light in the red region of the electromagnetic spectrum can traverse several layers of tissue. This is, for example, the reason why if we cover a flashlight with a hand, the backside of the hand looks red.

7.2.2 LIGHT ABSORPTION

Light absorption can occur at the energy level of electron orbitals or of chemical bond vibrations (stretching, bending, wiggling, etc.). This absorption of energy temporarily increases the total energy of the system (atom or molecule) and is followed by relaxation to the initial state. This relaxation releases the absorbed energy either as heat, in which case there is "loss" of the particular light package that was absorbed, or as reemission of light of different, usually lower, energy than that of the package originally absorbed.

Together with light scattering, strong light absorption is another limiting factor for light penetration into the tissue. Ultraviolet B (UVB) light, for example, does not penetrate further than the epidermal layers due to the strong absorption by proteins and DNA.

7.3 APPLICATIONS

In the case where light energy is fully transformed to heat, the loss of light energy can be detected and its magnitude is related to the concentration of the molecule responsible for the absorption (chromophore). The efficiency with which a molecule can absorb light this way is given by its absorption coefficient. This principle is used in DRS where the chromophore concentration is calculated from the intensity of absorption of a particular spectral band characteristic to the chromophore of interest. Evidently, DRS can be used in dermatokinetics when the drug can act as a chromophore itself or when it is affecting other skin chromophores; for example, a vasodilator would increase the concentration of hemoglobin, a known skin chromophore.

In the case when following absorption there is reemission of light energy, the emitted light can be detected and its amount is related to the concentration of the molecule responsible for the absorption and reemission (fluorophore). The efficiency with which a molecule can absorb and reemit light this way is given by its quantum yield. This principle is used in FS where the fluorophore concentration is calculated from the intensity of a particular spectral band characteristic to the molecule of interest and to the energy level of the incident light, which is related to its wavelength. Evidently, FS can be used in dermatokinetics when the drug itself is fluorescing or when it is affecting native skin fluorescence, for example, molecules such as α-hydroxy acids and their derivatives that can increase epidermal proliferation that is related to tryptophan fluorescence (Doukas et al. 2001, Bellemere et al. 2009).

The energy of mid-IR light is not enough to excite electrons to higher orbits, but is enough to accelerate certain vibrations of molecular bonds that are related to change in dipole moment. Therefore, molecules that have such types of bonds can absorb IR radiation, which can be shown as absorption bands in an IR spectrum. The position of these bands can be highly specific to a particular molecule such as a drug diffusing through the skin.

Finally, another type of spectroscopy that relates to vibrations of molecular bonds is Raman spectroscopy. The Raman phenomenon is a type of light scattering characterized not only by change in the direction of light travel but also by shifts in the light energy (relating to light frequency). Molecules with bonds that change their polarizability as they vibrate can be detected with this method. A Raman spectrum is expressed as a shift in the energy of the incident light, and, similarly to IR spectroscopy, the position of bands in a Raman spectrum can be highly specific to a molecule of interest.

We will now examine some specific applications of the above-mentioned spectroscopic techniques in the study of penetration of topically applied exogenous substances through the skin.

7.3.1 DIFFUSE REFLECTANCE SPECTROSCOPY IN DERMATOKINETICS

In DRS, the instrumentation includes a light source and a spectrometer (Figure 7.1). The two are connected to a bifurcated optical fiber bundle, the common end of which is the probe that comes in contact with the skin tissue. The optical fiber probe can have different configurations, including random or specified geometries for the illumination and collection fibers (those that connect to the light source and the detector correspondingly). The size of the individual fibers and the distance between the illumination and the collection fibers are important factors determining the tissue volume that is to be sampled. For shallow penetration, for example, when one is interested in the kinetics of drug penetration in the epidermis, small fibers (less than 100 μm) and short illumination–collection distances should be chosen. The opposite is true when one is

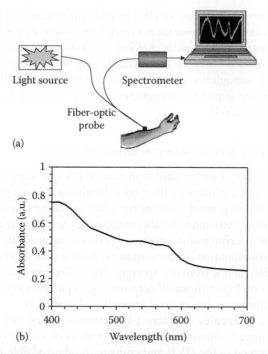

Light source   Spectrometer

Fiber-optic
probe

(a)

(b)

FIGURE 7.1 (a) (**See color insert.**) A diffuse reflectance spectroscopy (DRS) system comprises a light source, a probe, and a computer-controlled spectrometer. The probe can be a bifurcated optical fiber, as depicted here, or an integrating sphere. The geometry of the probe is one of the parameters determining the detection depth in the tissue. (b) Typical DRS spectra of untreated skin. The exponential-like decay toward longer wavelengths is attributed to melanin and the bands in the region 540–580 nm to hemoglobin.

interested to study events in the dermis. In that case, large distance of illumination to collection fiber is preferred.

Another alternative for a probe is an integrating sphere. This is a hollow sphere covered with a reflecting (white) material inside and with openings for light from the source, for a channel that goes to the detector and for a window to the sample surface. This type of probe allows for "diffuse" illumination, which means that the sample is illuminated from a wide range of angles. The same is true for the collection of remitted light form the tissue. Care needs to be taken to exclude the amount of secularly reflected light from the skin surface (Stamatas et al. 2008b).

Due to the fact that many drugs do not absorb in the visible part of the spectrum, the use of DRS in studying directly the drug concentration in the skin has been limited. An important drug that does absorb in the UV–short visible spectral range is 5-aminolevulinic acid (ALA). This compound has been widely used in the photodynamic therapy of non-melanoma skin cancers as a precursor to protoporphyrin IX (PpIX), a molecule that acts as a photosensitizer. Based on the absorption properties of ALA, its kinetics of penetration following topical application have been studied using DRS (Kim et al. 2007).

More frequent though is the use of DRS to study the effects of a topically applied substance. In this case, the penetration kinetics are inferred, for example, by the dynamics of vascular reactions. Kollias et al. have used this method to study the kinetics of SLS-induced irritation based on the spectral calculation of the apparent concentration of oxy and hemoglobin (Kollias et al. 1995). Similarly, the vasoconstrictive action of topically applied hydrogen peroxide can be documented using DRS (Stamatas and Kollias 2004).

7.3.2 Fluorescence Spectroscopy in Dermatokinetics

To perform FS in vivo, a setup similar to that of DRS is used, including a light source, a detector, and a bifurcated fiber optic bundle to act as a probe. The difference is that for FS it is important to use either a narrow band or monochromatic light for excitation (illumination) and to be able to spectrally analyze the emission (collection) with sufficient discrimination sensitivity between wavelengths. To this end, it is common to use as illumination a monochromatic laser source or a broadband source coupled to a double monochromator (prism). The collection end can either be an array of detectors or a high-efficiency detector (e.g., a photomultiplier tube) coupled to a double monochromator for spectral separation (Figure 7.2).

There are some molecules of interest in dermatokinetics that fluoresce themselves or can be linked to fluorescent moieties to act as beacons. One molecule that fluoresces when excited in the UVB and emits in the short visible range is salicylic acid and its skin penetration kinetics have been studied using FS (Stamatas et al. 2002). Many components of the skin can either fluoresce themselves in the spectral area where the molecule of interest is fluorescing or they may interfere by absorbing part of the excitation or the fluorescently emitted light. In such a case, care should be taken to correct for such artifacts. In the case of salicylic acid, a mathematical model was developed and a calibration curve was constructed so that the corrected measurement would be independent of the skin chromophore concentration (Stamatas et al. 2002).

Rhodes and Diffey reported the use of FS for the quantification of skin surface thickness of topically applied agents including a sunscreen, an antiseptic, and a steroid (Rhodes and Diffey 1997). The fluorescence intensity was found to strongly correlate to the logarithm of surface density allowing for the derivation of a formula for calculating an "equivalent thickness." Rate constants for the reduction of surface density could also be calculated.

As mentioned before, the kinetics of ALA as a photosensitizer precursor can be studied using DRS. More importantly, since the derivative molecule PpIX fluoresces upon excitation in the UV–short visible range, the course of generation of PpIX from its precursors can be documented in vivo using FS. Juzeniene et al. studied PpIX accumulation after topical application of either ALA or two of its ester derivatives in normal human skin (Juzeniene et al. 2006). For all three, drug fluorescence emission and excitation spectra exhibited similar spectral shapes indicating the formation of PpIX.

Using FS, researchers have found that the depth of PpIX production increased with increased application time of the hexyl ester of ALA (Zhao et al. 2006). In

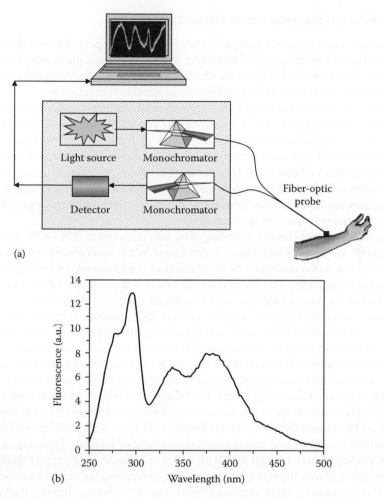

(a)

(b)

FIGURE 7.2 **(See color insert.)** (a) A fluorescence spectroscopy (FS) system comprises a light source, a probe, and a sensitive detector. Double monochromators after the light source and just before the detector allow for the selection of excitation and emission wavelengths, respectively. The probe is typically a bifurcated optical fiber. (b) A typical fluorescence spectrum of untreated skin acquired during a synchronous scan: both excitation and emission wavelengths β are changing by keeping a constant difference between the two (in this case $\Delta\lambda = 50\,nm$). Bands ascribed to tryptophan, collagen, and elastin cross-links are visible.

the same study, they were able to quantify pigmentary and vascular reactions in vivo using DRS.

Lesar et al. studied the application time of two different PpIX precursors on human skin in vivo and found that following application of either ALA or methyl amonolevulinate (MAL), PpIX fluorescence is dependent on the duration of application (Lesar et al. 2009). They also reported that ALA is more efficacious in producing PpIX compared to MAL.

7.3.3 INFRARED SPECTROSCOPY IN DERMATOKINETICS

The absorption spectrum of a sample in the IR part of the spectrum provides information on the vibrational states of molecular groups about specific bonds. This type of spectroscopy allows not only for the identification of molecular composition of the sample but also for information regarding the molecular organization.

The use of IR absorption spectroscopy was radically improved with the introduction of interferometers. These spectrometers record the pattern of constructive and destructive interference between two beams of light: the sample beam and a reference beam that does not interact with the sample. The Fourier transform of the interference pattern results in the IR absorption profile of the sample. Fourier transform infrared (FTIR) spectroscopy demonstrates reduced noise, increased sensitivity, more accurate frequency determination, and improved resolution compared to IR spectroscopy using dispersion spectrometers.

Since most biological tissues, including skin, are composed of 70%–80% water, the IR absorption spectrum of such tissues is dominated by the water absorption bands that can be broad due to the absorption by the plurality of conformations of hydrogen bonds between water molecules. This spectral dominance of water makes it difficult to obtain information about other molecular species in the IR spectrum of the tissue bulk. The top layers of skin comprising the stratum corneum (SC), however, contain significantly less water, with the concentration of water decreasing from the bulk of the tissue to the surface. This is also the place where most of the skin resistance to penetration of externally applied substances lies and therefore the SC is of great interest to dermatokinetics.

To isolate the information concerning the SC noninvasively, a method has developed that has been successfully coupled to IR spectroscopy. Attenuated total reflection (ATR) takes advantage of the minute loss of energy when an IR beam traveling through an IR-transparent solid crystal bounces off from well-polished surfaces of a crystal that come in close contact with the sample of interest. Typically, a zinc-selinide (ZnSe) crystal is used and with a flat or pointed geometry (Figure 7.3). A flat geometry allows for multiple bounces, thus increasing the signal-to-noise ratio (SNR), though limiting the in vivo use to flat areas of the skin, primarily the ventral side of the lower forearm. In contrast, the pointed geometry allows for coupling of the crystal to flexible optical fibers and therefore gives access to skin at any part of the body. Pointed geometry is limited to two optical bounces and therefore has lower SNR, but better spatial discrimination (small probed area) than the flat crystal.

At the point of reflection of light there is an evanescent electromagnetic field that extends beyond the surface of the crystal, penetrating the sample that comes in contact with the probe by approximately a quarter to a half of the wavelength of the IR beam. Applied to skin, the probed volume extends to a thickness of 1–2 μm covering practically about one layer of corneocytes in the SC. This part of the SC absorbs IR energy at frequencies corresponding to its normal absorption spectrum.

The absorption signals in ATR-FTIR are weak and therefore sensitive detectors are used to record the signals. The liquid-nitrogen-cooled mercury-cadmium-telluride detector is commonly used for this purpose.

Since the IR absorption signals depend on the degree of contact between the crystal and the skin, in order to quantitatively compare absorption bands, the band of

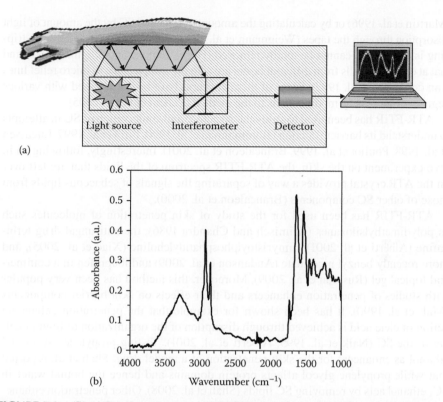

(a)

(b)

FIGURE 7.3 **(See color insert.)** (a) An ATR-FTIR spectroscopy system comprises a laser light source, an internal reflectance element (typically ZnSe crystal) probe, and an interferometer with a liquid-nitrogen-cooled detector. (b) A typical IR absorption spectrum of untreated skin showing the long wave number (2400–4000 cm^{-1}) and the fingerprint (1000–2000 cm^{-1}) regions in a single scan. One can observe the OH stretch, CH$_2$ asymmetric and symmetric stretches, CO stretch, and amide I, amide II, and CH$_2$ scissoring modes.

interest needs to be normalized to another signal that is similarly dependent on the contact. The amide I band is commonly used for this purpose.

The shallow penetration is an advantage and a limitation for the application of the method in dermatokinetics. The advantage is a good resolution in the z-direction (depth) and the limitation is that with each experiment we get only the surface information. Therefore, if one is interested to study the concentration profile of a drug in the SC, the ATR-FTIR method needs to be coupled to sequential stripping of the SC layers. This can be achieved with sequential application of adhesive tapes alternated with ATR-FTIR measurements. This technique, though not completely noninvasive and definitely not nondestructive (i.e., one cannot measure at the same spot more than once), has been widely used in the literature (see references in the following two paragraphs). The corneocytes at each depth may not only have varying adhesion strengths among them but also varying qualities that influence their adhesion to the tape. Since each adhesive tape does not remove exactly the same amount of SC, care must be taken to correct the results accordingly. This can be accomplished by weighing the tapes before and after stripping

(Marttin et al. 1996) or by calculating the amount of removed SC by the amount of light absorption through the tapes (Weigmann et al. 1999). Another limitation of tape stripping is that the SC cannot be totally removed (Hojyo-Tomoka and Kligman 1972) and that at each tape, cells from different layers are actually sampled due to micro relief lines (van der Molen et al. 1997). Some of these concerns have been addressed with various degrees of success by modifications to the method (Lademann et al. 2005).

ATR-FTIR has been used to study the structure and composition of SC in attempts to understand its barrier function (Bommannan et al. 1990, Pirot et al. 1997, Lucassen et al. 1998, Pouliot et al. 1999, Brancaleon et al. 2001). Interestingly, following an in vivo experiment on the skin, the ATR-FTIR spectrum of the lipids that are left over on the ATR crystal provides a way of separating the signals of sebaceous lipids from those of other SC components (Brancaleon et al. 2000).

ATR-FTIR has been used for the study of skin penetration of molecules such as polydimethylsiloxanes (Klimisch and Chandra 1986), the antifungal drug terbinafine (Alberti et al. 2001), dimyristoylphosphatidylcholine (Xiao et al. 2005), and more recently benzyl nicotinate (Andanson et al. 2009) and ibuprofen in a commercial topical gel (Russeau et al. 2009). Moreover, this method has been very popular with studies of penetration enhancers and their effects on skin barrier components (Mak et al. 1990). It has been shown for example that the penetration-enhancing action of oleic acid is achieved through disruption of the organization of lipid lamellae in the SC (Naik et al. 1995, Alberti et al. 2001). Testing propylene glycol and ethanol as enhancers for the skin penetration of a model drug, Shah et al. reported that while propylene glycol affects protein domains and hence the bound water in SC, ethanol acts by removing SC lipids (Shah et al. 2008). Other penetration enhancers studied using FTIR include ethanol (Bommannan et al. 1991, Andanson et al. 2009), urea (Alberti et al. 2001), 2-pyrrolidone (Alberti et al. 2001), dimethylsulfoxide (Mendelsohn et al. 2006, Jiang et al. 2008), and techniques such as iontophoresis (Curdy et al. 2001) and the use of liposomes (Mendelsohn et al. 2006).

7.3.4 RAMAN SPECTROSCOPY IN DERMATOKINETICS

Most work on transdermal delivery, including the ATR-FTIR works mentioned above, implicates protocols that are either ex vivo or involve tape stripping with their own disadvantages. The Holy Grail of dermatokinetics remains in vivo quantitative monitoring of active concentration through the skin layers. The method that approaches very close to this ideal is in vivo confocal Raman microspectroscopy, a technique that has met with tremendous progress the past decade.

Similar to IR, Raman spectroscopy gives information about the presence and intensity of vibrational modes between chemical bonds. Whereas IR-active molecular bonds are those that change their dipole moment upon absorbing electromagnetic radiation of certain energy, Raman-active bonds are those that change their polarizability. Furthermore, in contrast to IR absorption where the absorbed light energy is being transformed to heat, the Raman phenomenon is considered as "inelastic scattering," which indicates that part of the absorbed light is reemitted at a different energy (wavelength or frequency). Raman scattering is inherently very weak and therefore specialized cooled detectors are required.

Although Raman spectroscopy can be used in combination with an ATR probe (Futamata et al. 1994), to our knowledge the ATR-Raman method has not been applied to skin penetration studies. Instead, coupling of Raman spectroscopy to a confocal arrangement (Figure 7.4) has been much more popular in skin research, since it allows for noninvasive calculations of concentration profiles through the SC, thus avoiding the need for tape stripping.

Typically, a laser in the red or short IR range is used as a light source. The choice for the laser wavelength is limited by skin fluorescence, which in Raman spectroscopy gives a background that needs to be corrected, and the achievable penetration depth, which is in turn limited by the absorption and scattering coefficients of the tissue.

Caspers et al. were the first to describe a confocal Raman microspectroscopy system for in vivo studies of human skin (Caspers et al. 1998). Since then, much

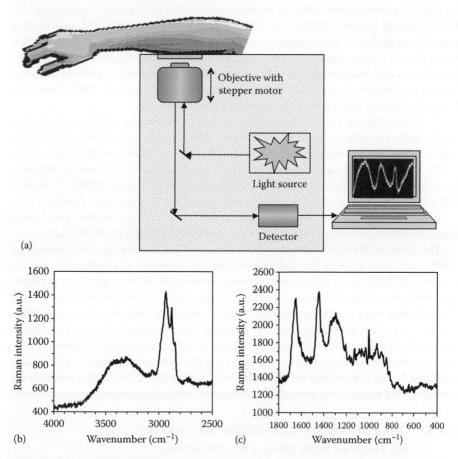

FIGURE 7.4 (See color insert.) (a) A confocal Raman microspectroscopy system comprises a laser light source, a microscope objective, the position of which is controlled by a z-axis stepper motor, and a sensitive detector. (b) A typical Raman spectrum of untreated skin showing the long wavenumber (2400–4000 cm^{-1}) and (c) the fingerprint (400–2400 cm^{-1}) regions.

work has been focused toward understanding skin components that would give Raman signals (Hata et al. 2000, Caspers et al. 2001, Egawa and Tagami 2008). Most interestingly, water concentration profiles through the SC can be constructed based on Raman spectra (Caspers et al. 2001, Chrit et al. 2006, Egawa et al. 2007) with obvious consequences for penetration of exogenous hydrophilic molecules. As with ATR-FTIR spectroscopy, lipid and protein organization can be inferred from Raman spectra and in particular from the position and the shape of bands that correspond to CH stretching and bending modes for lipids and the amide stretching modes for proteins.

More importantly though, the concentrations of Raman-active molecules that have been topically applied on the skin can be monitored and one can study the evolution in time of concentration profiles. Using in vivo confocal Raman the penetration of *trans*-retinol, a cosmetic active, has been reported (Pudney et al. 2007) as well as a comparison between retinol-containing formulations representing different delivery systems (Melot et al. 2009). Other examples of use of Raman include penetration studies of metroconazole, a drug for rosacea (Tfayli et al. 2007), and titanium dioxide nanoparticles in a sunscreen formulation (Lademann et al. 1999). The penetration and mechanisms of action of permeation enhancers, such as dimethyl sulfoxide (Caspers et al. 2002) and iminosulfuranes (Song et al. 2005), have also been investigated using confocal Raman microspectroscopy.

Molecules that are either not active in Raman or overlapping with bands from skin components are inherently more difficult to monitor. The penetration of topically applied lipids, for example, is difficult in the "fingerprint" wave number region (400–2400 cm^{-1}), due to the presence of native skin lipids. Stamatas et al., however, were able to monitor lipid penetration using the "high wave number" region (2400–4000 cm^{-1}) and subtracting the contribution of lipids native to skin (Stamatas et al. 2008a).

In ex vivo experiments, Raman spectroscopy has been used to study penetration and even chemical transformation of drugs in the skin (Zhang et al. 2007a,b).

The confocal Raman method has been criticized for two reasons (Herkenne et al. 2008): (a) it requires that the molecule of interest is present at an adequate concentration and should have a sufficient quantum yield in the Raman spectrum to permit differentiation from the skin molecules and (b) only relative concentrations rather than absolute ones can be determined, as is for example with extraction and HPLC analysis. The first problem is present with any other method. One needs to select the method most appropriate for the molecule of interest and this depends on what signals (absorption, fluorescence, Raman, etc.) the molecule possesses. The second criticism is also common in all methods since even with the extraction methods the concentration needs to be normalized to the total amount of the extracted material.

7.3.5   OTHER METHODS: PHOTOACOUSTIC SPECTROSCOPY
AND HIGHER-ORDER OPTICAL SPECTROSCOPIES

Photothermal or photoacoustic spectroscopy (PAS) is another nondestructive method that has been proposed for the study of transdermal drug penetration (Kölmel et al. 1986). A modulated monochromatic beam is projected upon the skin and causes periodic heating due to light absorbance, which in turn results in pressure oscillations that can be detected by a sensitive microphone. The measurement depth is controlled by changing the modulation frequency of the light beam incident to the skin surface. Due to low spatial resolution and other limitations of the photoacoustic cells, the method has been almost exclusively limited to in vitro or ex vivo systems (Hanh et al. 2000).

Higher-order (or nonlinear) spectroscopies include second harmonic generation (SHG), coherent anti-Stokes Raman (CARS), and multiphoton fluorescence (MPF). They provide high spatial resolution for microscopic imaging, decreased photodamage compared to FS, and penetration depths up to 1000 µm (Hanson and Bardeen 2009). Both epidermal and dermal structures can be observed: MPF can detect epidermal fluorescence attributed to keratin, NAD(P)H, flavins, and melanin and dermal signals from collagen and elastin fluorescence; CARS can be used to visualize skin lipids; and SHG can detect signals from collagen in the dermis (Schenke-Layland et al. 2006, Palero et al. 2007). Structural properties of the SC have been correlated with transport rates of fluorescing reporter molecules such as lipophylic and hydrophilic analogues of rhodamine B (Yu et al. 2003). One study reported the evolution of drug penetration in human skin in vivo (König et al. 2006), whereas the effects of the penetration enhancer oleic acid on transdermal transport have been studied in vitro (Yu et al. 2001). The major hurdles for in vivo applications of such techniques are the long time of acquisition and the need for stabilizing the skin site under investigation to avoid motion artifacts.

7.4 SUMMARY

Spectroscopic methods provide rapid noninvasive means to dynamically study skin penetration kinetics of topically applied substances. Such substances can be monitored directly (following a spectroscopic signal of the substance of interest) or indirectly (following signals related to changes in the molecules native to skin inferred by the exogenous substance).

Current technology provides the investigators with an arsenal of various spectroscopic methods based on reflectance, fluorescence, absorption in the IR, or Raman scattering. The advantages and limitations of these methods are reviewed in Table 7.1.

DRS is a cost-effective method for monitoring drug penetration, but the signals can be nonspecific. It may be more useful as a tool to monitor vascular or pigmentary effects of the applied substances.

Like DRS, fluorescence-based techniques can be used both for monitoring the effects of actives on the skin fluorophores or monitoring directly the active penetration if the molecule of interest fluoresces. If it does not fluoresce itself, a fluorescence tracer can be used to tag the active.

ATR-FTIR spectroscopy provides more specific signals than DRS or FS. However, although this method has been used more widely than the others, it comes with its own shortcomings. In order for it to be applied in dermatokinetics it needs to be combined with tape stripping the skin. It is, therefore, plagued with all the limitations of stripping such as nonuniform removal of SC and variable adhesive power.

TABLE 7.1

Comparison of In Vivo Spectroscopic Methods for the Study of Transdermal Kinetics of Topically Applied Active Compounds

Method	Advantages	Disadvantages
DRS	Very portable; very rapid; inexpensive; ideal for the studying the effects of the active on the vascular and pigmentary systems	Not many compounds have absorption in the UV, visible, or IR range; signals may be nonspecific and analysis may be difficult; no depth profiling without tape stripping
FS	Portable; rapid; can be used to study the effects of the active on epidermal cell proliferation; can be used in a confocal arrangement for depth profiling	Not many compounds fluoresce; fluorescence bands may be nonspecific; surrogate indicators may need to be used, of which only few are FDA approved
ATR-FTIR	Rapid; very specific signal; shallow penetration resulting in good axial resolution	No depth profiling without tape stripping
Raman	Rapid; very specific signal; depth profiling can be achieved with a confocal arrangement; very good axial resolution	Comparatively more expensive; difficult to study penetration beyond epidermis (though not impossible)

In vivo confocal Raman microspectroscopy is the method that comes closer to the ideal of nondestructive quantitative monitoring of transdermal penetration kinetics.

PAS and nonlinear spectroscopies have shown a high potential in ex vivo studies, but have so far have found only limited application for in vivo monitoring of transdermal active penetration.

Regarding the future research in this field, more validation studies are needed. This is not an easy task as the only methods to use as "golden" standards are invasive and each with their own shortcomings. Finally, studies that compare results observed by two or more of the in vivo noninvasive methods are needed to understand their differences and evaluate their accuracy.

BIBLIOGRAPHY

Alberti I., Kalia Y. N., Naik A., Bonny J.-D., Guy R. H. 2001. In vivo assessment of enhanced topical delivery of terbinafine to human stratum corneum. *J Control Release* 71:319–27.

Andanson J. M., Hadgraft J., Kazarian S. G. 2009. In situ permeation study of drug through the stratum corneum using attenuated total reflection fourier transform infrared spectroscopic imaging. *J Biomed Opt* 14:034011.

Bellemere G., Stamatas G. N., Bruere V., Bertin C., Issachar N., Oddos T. 2009. Antiaging action of retinol: From molecular to clinical. *Skin Pharmacol Physiol* 22:200–209.

Bommannan D., Potts R. O., Guy R. H. 1990. Examination of stratum corneum barrier function in vivo by infrared spectroscopy. *J Invest Dermatol* 95:403–408.

Bommannan D., Potts R. O., Guy R. H. 1991. Examination of the effect of ethanol on human stratum corneum in vivo using infrared spectroscopy. *J Control Release* 16:299–304.

Brancaleon L., Bamberg M. P., Kollias N. 2000. Spectral differences between stratum corneum and sebaceous molecular components in the mid-IR. *Appl Spectrosc* 54:1175–1182.

Brancaleon L., Durkin A. J., Tu J. H., Menaker G., Fallon J. D., Kollias N. 2001. In vivo fluorescence spectroscopy of nonmelanoma skin cancer. *Photochem Photobiol* 73:178–183.

Caspers P. J., Lucassen G. W., Carter E. A., Bruining H. A., Puppels G. J. 2001. In vivo confocal Raman microspectroscopy of the skin: Noninvasive determination of molecular concentration profiles. *J Invest Dermatol* 116:434–442.

Caspers P. J., Lucassen G. W., Wolthuis R., Bruining H. A., Puppels G. J. 1998. In vitro and in vivo Raman spectroscopy of human skin. *Biospectroscopy* 4:S31–S39.

Caspers P. J., Williams A. C., Carter E. A., Edwards H. G., Barry B. W., Bruining H. A., Puppels G. J. 2002. Monitoring the penetration enhancer dimethyl sulfoxide in human stratum corneum in vivo by confocal Raman spectroscopy. *Pharm Res* 19:1577–1580.

Chrit L., Bastien P., Sockalingum G. D., Batisse D., Leroy F., Manfait M., Hadjur C. 2006. An in vivo randomized study of human skin moisturization by a new confocal Raman fiberoptic microprobe: Assessment of a glycerol-based hydration cream. *Skin Pharmacol Physiol* 19:207–215.

Curdy C., Kalia Y. N., Guy R. H. 2001. Non-invasive assessment of the effects of iontophoresis on human skin in-vivo. *J Pharm Pharmacol* 53:769–777.

Doukas A. G., Soukos N. S., Babusis S., Appa Y., Kollias N. 2001. Fluorescence excitation spectroscopy for the measurement of epidermal proliferation. *Photochem Photobiol* 74:96–102.

Egawa M., Hirao T., Takahashi M. 2007. In vivo estimation of stratum corneum thickness from water concentration profiles obtained with Raman spectroscopy. *Acta Derm Venereol* 87:4–8.

Egawa M. and Tagami H. 2008. Comparison of the depth profiles of water and water-binding substances in the stratum corneum determined in vivo by Raman spectroscopy between the cheek and volar forearm skin: Effects of age, seasonal changes and artificial forced hydration. *Br J Dermatol* 158:251–260.

Escobar-Chavez J. J., Merino-Sanjuan V., Lopez-Cervantes M., Urban-Morlan Z., Pinon-Segundo E., Quintanar-Guerrero D., Ganem-Quintanar A. 2008. The tape-stripping technique as a method for drug quantification in skin. *J Pharm Pharm Sci* 11:104–130.

Futamata M., Borthen P., Thomassen J., Schumacher D., Otto A. 1994. Application of an ATR method in raman spectroscopy. *Appl Spectrosc* 48:252–260.

Hanh B. D., Neubert R. H., Wartewig S., Christ A., Hentzch C. 2000. Drug penetration as studied by noninvasive methods: Fourier transform infrared-attenuated total reflection, fourier transform infrared, and ultraviolet photoacoustic spectroscopy. *J Pharm Sci* 89:1106–1113.

Hanson K. M. and Bardeen C. J. 2009. Application of nonlinear optical microscopy for imaging skin. *Photochem Photobiol* 85:33–44.

Hata T. R., Scholz T. A., Ermakov I. V., McClane R. W., Khachik F., Gellermann W., Pershing L. K. 2000. Non-invasive Raman spectroscopic detection of carotenoids in human skin. *J Invest Dermatol* 115:441–448.

Herkenne C., Alberti I., Naik A., Kalia Y. N., Mathy F. X., Preat V., Guy R. H. 2008. In vivo methods for the assessment of topical drug bioavailability. *Pharm Res* 25:87–103.

Hojyo-Tomoka M. T. and Kligman A. M. 1972. Does cellophane tape stripping remove the horny layer? *Arch Dermatol* 106:767–768.

Jiang J., Boese M., Turner P., Wang R. K. 2008. Penetration kinetics of dimethyl sulphoxide and glycerol in dynamic optical clearing of porcine skin tissue in vitro studied by fourier transform infrared spectroscopic imaging. *J Biomed Opt* 13:021105.

Juzeniene A., Juzenas P., Ma L. W., Iani V., Moan J. 2006. Topical application of 5-aminolaevulinic acid, methyl 5-aminolaevulinate and hexyl 5-aminolaevulinate on normal human skin. *Br J Dermatol* 155:791–799.

Kim K. H., Jheon S., Kim J. K. 2007. In vivo skin absorption dynamics of topically applied pharmaceuticals monitored by fiber-optic diffuse reflectance spectroscopy. *Spectrochim Acta A Mol Biomol Spectrosc* 66:768–772.

Klimisch H. M. and Chandra G. 1986. Use of fourier transform infrared spectroscopy with attenuated total reflectance for in vivo quantitation of polydimethylsiloxanes on human skin. *J Soc Cosmet Chem* 37:73–87.

Kollias N., Gillies R., Muccini J. A., Uyeyama R. K., Phillips S. B., Drake L. A. 1995. A single parameter, oxygenated hemoglobin, can be used to quantify experimental irritant-induced inflammation. *J Invest Dermatol* 104:421–424.

Kölmel K., Nikolaus A., Sennhenn B., Giese K. 1986. Evaluation of drug penetration into the skin by photoacoustic measurement. *J Soc Cosmet Chem* 37:375–385.

König K., Ehlers A., Stracke F., Riemann I. 2006. In vivo drug screening in human skin using femptosecond laser multiphoton tomography. *Skin Pharmacol Physiol* 19:78–88.

Lademann J., Weigmann H., Rickmeyer C., Barthelmes H., Schaefer H., Mueller G., Sterry W. 1999. Penetration of titanium dioxide microparticles in a sunscreen formulation into the horny layer and the follicular orifice. *Skin Pharmacol Appl Skin Physiol* 12:247–256.

Lademann J., Weigmann H. J., Schanzer S., Richter H., Audring H., Antoniou C., Tsikrikas G., Gers-Barlag H., Sterry W. 2005. Optical investigations to avoid the disturbing influences of furrows and wrinkles quantifying penetration of drugs and cosmetics into the skin by tape stripping. *J Biomed Opt* 10:054015.

Lesar A., Ferguson J., Moseley H. 2009. A time course investigation of the fluorescence induced by topical application of 5-aminolevulinic acid and methyl aminolevulinate on normal human skin. *Photodermatol Photoimmunol Photomed* 25:191–195.

Lucassen G. W., van Veen G. N. A., J. J. J. A. 1998. Band analysis of hydrated human skin stratum corneum attenuated total reflectance fourier transform infrared spectra in vivo. *J Biomed Opt* 3:267–280.

Mak V. H., Potts R. O., Guy R. H. 1990. Percutaneous penetration enhancement in vivo measured by attenuated total reflectance infrared spectroscopy. *Pharm Res* 7:835–841.

Marttin E., Neelissen-Subnel M. T., De Haan F. H., Bodde H. E. 1996. A critical comparison of methods to quantify stratum corneum removed by tape stripping. *Skin Pharmacol* 9:69–77.

Melot M., Pudney P. D., Williamson A. M., Caspers P. J., van der Pol A., Puppels G. J. 2009. Studying the effectiveness of penetration enhancers to deliver retinol through the stratum cornum by in vivo confocal Raman spectroscopy. *J Control Release* 138:32–39.

Mendelsohn R., Flach C. R., Moore D. J. 2006. Determination of molecular conformation and permeation in skin via ir spectroscopy, microscopy, and imaging. *Biochim Biophys Acta* 1758:923–933.

Naik A., Pechtold L. A. R. M., Potts R. O., Guy R. H. 1995. Mechanism of oleic acid-induced skin penetration enhancement in vivo in humans. *J Control Release* 37:299–306.

Palero J. A., de Bruijn H. S., van der Ploeg van der Heuvel A., Sterenborg H. J. C. M., Gerritsen H. C. 2007. Spectrally resolved multiphoton imaging of in vivo and excised mouse skin tissues. *Biophys J* 93:992–1007.

Pirot F., Kalia Y. N., Stinchcomb A. L., Keating G., Bunge A., Guy R. H. 1997. Characterization of the permeability barrier of human skin in vivo. *Proc Natl Acad Sci U S A* 94:1562–1567.

Pouliot R., Germain L., Auger F. A., Tremblay N., Juhasz J. 1999. Physical characterization of the stratum corneum of an in vitro human skin equivalent produced by tissue engineering and its comparison with normal human skin by ATR-FTIR spectroscopy and thermal analysis (dsc). *Biochim Biophys Acta* 1439:341–352.

Pudney P. D., Melot M., Caspers P. J., van der Pol A., Puppels G. J. 2007. An in vivo confocal Raman study of the delivery of trans retinol to the skin. *Appl Spectrosc* 61:804–811.

Rhodes L. E. and Diffey B. L. 1997. Fluorescence spectroscopy: A rapid, noninvasive method for measurement of skin surface thickness of topical agents. *Br J Dermatol* 136:12–17.

Russeau W., Mitchell J., Tetteh J., Lane M. E., Hadgraft J. 2009. Investigation of the permeation of model formulations and a commercial ibuprofen formulation in carbosil and human skin using ATR-FTIR and multivariate spectral analysis. *Int J Pharm* 374:17–25.

Schenke-Layland K., Riemann I., Damour O., Stock U. A., König K. 2006. Two-photon microscopes and in vivo multiphoton tomographs—Powerful diagnostic tools for tissue engineering and drug delivery. *Adv Drug Delivery Rev* 58:878–896.

Shah D. K., Khandavilli S., Panchagnula R. 2008. Alteration of skin hydration and its barrier function by vehicle and permeation enhancers: A study using TGA, FTIR, TEWL and drug permeation as markers. *Methods Find Exp Clin Pharmacol* 30:499–512.

Song Y., Xiao C., Mendelsohn R., Zheng T., Strekowski L., Michniak B. 2005. Investigation of iminosulfuranes as novel transdermal penetration enhancers: Enhancement activity and cytotoxicity. *Pharm Res* 22:1918–1925.

Stamatas G. N., de Sterke J., Hauser M., von Stetten O., van der Pol A. 2008a. Lipid uptake and skin occlusion following topical application of oils on adult and infant skin. *J Dermatol Sci* 50:135–142.

Stamatas G. N. and Kollias N. 2004. Blood stasis contributions to the perception of skin pigmentation. *J Biomed Opt* 9:315–322.

Stamatas G. N., Wu J., Kollias N. 2002. Non-invasive method for quantitative evaluation of exogenous compound deposition on skin. *J Invest Dermatol* 118:295–302.

Stamatas G. N., Zmudzka B. Z., Kollias N., Beer J. Z. 2008b. In vivo measurement of skin erythema and pigmentation: New means of implementation of diffuse reflectance spectroscopy with a commercial instrument. *Br J Dermatol* 159:683–690.

Tfayli A., Piot O., Pitre F., Manfait M. 2007. Follow-up of drug permeation through excised human skin with confocal Raman microspectroscopy. *Eur Biophys J* 36:1049–1058.

van der Molen R. G., Spies F., van 't Noordende J. M., Boelsma E., Mommaas A. M., Koerten H. K. 1997. Tape stripping of human stratum corneum yields cell layers that originate from various depths because of furrows in the skin. *Arch Dermatol Res* 289:514–518.

Weigmann H., Lademann J., Meffert H., Schaefer H., Sterry W. 1999. Determination of the horny layer profile by tape stripping in combination with optical spectroscopy in the visible range as a prerequisite to quantify percutaneous absorption. *Skin Pharmacol Appl Skin Physiol* 12:34–45.

Xiao C., Moore D. J., Flach C. R., Mendelsohn R. 2005. Permeation of dimyristoylphosphatidylcholine into skin—Structural and spatial information from IR and Raman microscopic imaging, vibrational spectroscopy. *Vibr Spectrosc* 38:151–158.

Yu B., Dong C. Y., So P. T. C., Blankschtein D., Langer R. 2001. In vitro visualization and quantification of oleic acid induced changes in transdermal transport using two-photon fluorescence microscopy. *J Invest Dermatol* 117:16–25.

Yu B., Kim K. H., So P. T. C., Blankschtein D., Langer R. 2003. Evaluation of fluorescent probe surface intensities as an indicator of transdermal permeant distributions using wide-area two-photon fluorescence microsocpy. *J Pharm Sci* 92:2354–2365.

Zhang G., Flach C. R., Mendelsohn R. 2007a. Tracking the dephosphorylation of resveratrol triphosphate in skin by confocal Raman microscopy. *J Control Release* 123:141–147.

Zhang G., Moore D. J., Sloan K. B., Flach C. R., Mendelsohn R. 2007b. Imaging the prodrug-to-drug transformation of a 5-fluorouracil derivative in skin by confocal Raman microscopy. *J Invest Dermatol* 127:1205–1209.

Zhao L., Nielsen K. P., Juzeniene A., Juzenas P., Lani V., Ma L. W., Stamnes K., Stamnes J. J., Moan J. 2006. Spectroscopic measurements of photoinduced processes in human skin after topical application of the hexyl ester of 5-aminolevulinic acid. *J Environ Pathol Toxicol Oncol* 25:307–320.

8 Regulatory Perspective of Dermatokinetic Studies*

April C. Braddy and Dale P. Conner

CONTENTS

In this chapter, the U.S. Food and Drug Administration (U.S.-FDA) approval process for generic topical dermatological drug products is discussed. The approval of drug products is governed by regulations and guidances that facilitate the development of bioequivalent drug products. The goal is for these generic products to be pharmaceutically and therapeutically equivalent to the reference-listed drug product. The dosage form and indication of the drug product is a significant determining factor as to which scientific/clinical method, such as waivers of in vivo studies, in vivo pharmacokinetic studies, in vivo pharmacodynamic studies, bioequivalence trials with clinical endpoints, and/or in vitro studies, is the most suitable to demonstrate bioequivalence for a particular topical dermatological drug product. Also discussed in this chapter are some of the previous and current studies of scientific and/or clinical methods that provide the current basis for the U.S.-FDA recommendations.

8.1 SKIN BIOAVAILABILITY AND BIOEQUIVALENCE STUDIES

The U.S.-FDA approval process for a generic drug product is based, in part, on an applicant establishing bioequivalence between the reference listed drug (RLD), usually the original innovator drug product, and the proposed generic drug product. In 1984, the Drug Price Competition and Patent Term Restoration Act (Waxman-Hatch Act) was passed by Congress (United States Code, 1984). This law made dramatic changes to

---

* Disclaimer: Views expressed in this chapter are those of the authors and not necessarily of the U.S.-FDA.

the Federal Food, Drug, and Cosmetic Act (FFDCA) by permitting the approval of generic drug products through the submission and subsequent approval of Abbreviated New Drug Applications (ANDA) referencing innovator drug products approved after 1962. Generic drug products are pharmaceutically equivalent to the RLD, meaning that they contain the same amount of the active drug and in the same dosage form as the RLD and meet the same compendial or other applicable standards (i.e., strength, quality, purity, and identity). Bioequivalence studies serve to confirm that the generic product has equivalent therapeutic efficacy and safety to the approved RLD. These studies accomplish this by showing that the proposed generic drug product delivers the drug to the sites of therapeutic activity in the body at the same rate and extent as the RLD.

For topical dermatological drug products to be considered therapeutically equivalent, a test drug product must be pharmaceutically equivalent and also bioequivalent to the RLD. According to the U.S. Code of Federal Regulations (CFR), 21 CFR § 320.24, bioavailability may be measured and/or bioequivalence may be established based on the following (in descending order):

- In vivo pharmacokinetic tests in humans in which the active moiety is measured in biological fluids (i.e., whole blood, serum, plasma, or urine) as a function time
- In vivo pharmacodynamic tests in human in which the pharmacological effect of the drug product can be measured as a function time
- Well-controlled clinical trials to establish safety or efficacy between the two drug products
- An in vitro test approach currently acceptable to the U.S.-FDA

For topical dermatological solutions, in most cases, bioequivalence is considered self-evident when the components of the innovator and generic drug products are qualitatively and quantitatively the same (±5%). In this instance, a waiver of in vivo bioequivalence studies can be granted as stated in 21 CFR § 320.21 (b).

In general, in vivo pharmacokinetic studies are not feasible for topical dermatological drug products due to the fact that (1) many topical dermatological drug products do not consistently produce measureable concentrations in biological fluids such as plasma and (2) drug concentrations in plasma do not represent drug levels at the site of action (only represents concentration after passage through the target compartment). The absence of appropriate pharmacodynamic effects that can be used to evaluate drug availability at the site of activity is also another hurdle in the development of methods to assess bioavailability and/or bioequivalence for topical dermatological drug products. Based on this information, the U.S.-FDA has typically relied on bioequivalence studies with clinical endpoints, except in the case of topical corticosteroids in which in vivo pharmacodynamic bioequivalence studies are recommended.

8.1.1 BIOEQUIVALENCE STUDIES WITH CLINICAL ENDPOINTS

Bioequivalence studies with clinical endpoints allow for the establishment of bioequivalence between two drug products based on the demonstration of equivalent safety and efficacy in patients. However, these particular studies are expensive, time-consuming, labor-intensive, require a large study population, and sometimes give results that are difficult to analyze statistically.

8.1.2   In Vivo Pharmacodynamic Model

The primary example of a pharmacodynamic approach is based on the ability of topical dermatological corticosteroids to produce a visible blanching response as a result of vasoconstriction of the skin microvasculature (Stoughton 1987, 1992). Based on the U.S.-FDA Guidance for Industry: Guidance Topical Dermatologic Corticosteroids: In Vivo Bioequivalence (issued June 2, 1995), the Stoughton–McKenzie vasoconstrictor bioassay is used to assess the bioequivalence of topical corticosteroids by utilizing a study in two parts. The guidance recommends an initial pilot dose–duration response study that is used to determine the appropriate dose duration for use in a subsequent pivotal study. Data analysis of the vasoconstrictor response is based on an Emax model (population modeling), (Singh et al. 1999). The pilot dose–duration response study to determine an initial estimate of Emax for the product is conducted only on the RLD product. The pivotal pharmacodynamic study incorporates a replicate design and documentation of acceptable individual subject dose–duration response by comparing the test and RLD products. Based on this model, the test and RLD products should meet the acceptable bioequivalence criteria of 80.00%–125.00% for the observed vasoconstrictor response. Healthy subjects are enrolled in this study.

This method is a more direct and efficient indicator of local bioequivalence and is inexpensive as compared to bioequivalence studies with clinical endpoints. Unfortunately, this method is only applicable to topical corticosteroids.

8.1.3   In Vitro Release Test

The U.S.-FDA Guidance for Industry: Nonsterile Semisolid Dosage Forms, Scale-Up and Post-Approval Changes: Chemistry, Manufacturing, and Control; In Vitro Release Testing and In Vivo Bioequivalence Documentation (issued May 1997) addresses nonsterile topical preparations such as creams, gels, lotions, and ointments. It provides recommendations for in vitro release tests and/or in vivo bioequivalence tests to support changes to (1) the components or composition, (2) the manufacturing (process and equipment), (3) scale-up/scale-down of manufacture, and/or (4) the site of manufacture of a semisolid formulation during the postapproval period. The in vitro release method for topical dosage forms is based on an open chamber diffusion cell system such as a Franz cell system, fitted usually with a synthetic membrane. In most cases, in vitro release tests are very useful to assess product (formulation) sameness between pre-change and post-change products. The in vitro release rate tests should not be used as a surrogate or replacement for in vivo bioavailability or bioequivalence studies. The development and validation of in vitro release tests are not required for approval of drug applications (New Drug Applications [NDAs] or ANDAs). Also is not required as a routine batch-to-batch quality control test.

8.2   U.S.-FDA Guidelines for Dermatokinetic Studies

The dermatokinetic method was proposed as a method for confirming the bioequivalence of topical dermatological products. Much of the drug absorbed into the skin from topical products must pass through the stratum corneum (outer layer of the skin). For many chemical compounds, this thin outer layer of the skin forms a rate-limiting barrier. In the dermatokinetic method, the rate of drug penetration from a product through this outer layer was evaluated by applying the product to an area of skin and later "striping" off the layers of the stratum corneum with tape and evaluating the kinetics of drug penetration. Some of the previously claimed advantages of the dermatokinetic method of skin (tape) stripping are its applicability to all classes of topical dermatological drug products, ability to measure the drug concentration near the vicinity of the site of action, and that assessment of the drug concentration in the stratum corneum may parallel the concentration in deeper skin layers, such as the epidermis and dermis (Shah 2001). However, the perceived disadvantages in the dermatokinetic method are due to the lack of established clinical relevancy, the possible inability of the method to differentiate between two topical drug product formulations and multiple strengths, and the reliability and reproducibility of the method between different laboratories. The application of the dermatokinetic method raised concerns based on (1) the use of the stratum corneum to determine bioequivalence, (2) the use of healthy versus diseased skin, (3) furrows in the skin, and (4) hair follicles and follicular penetration. Each one of these will be discussed briefly.

With the proposed use of stratum corneum drug concentrations over time to determine bioequivalence, it is suggested that if both topical dermatological drug products exhibit similar dermatokinetic profiles in the stratum corneum, then both should have the same delivery to the sites of activity within the skin and therefore the same safety and efficacy profile. This is based on the assumption that in dermatokinetic studies the comparison is between two similar topical dermatological drug product formulations; therefore, the only possible difference may be in the partitioning of the drug from the product vehicle into the stratum corneum that is the rate-limiting step in local drug absorption. It has been postulated that if the stratum corneum is damaged and/or diseased, it may reduce the barrier to drug absorption allowing for a greater and more rapid delivery of the drug to the site of action with a much less apparent difference between different products. Therefore, the intact stratum corneum in healthy subjects may be a more rigorous and sensitive test of comparative drug absorption via the percutaneous route than the diseased skin.

It is known that the skin surface is not flat but contains furrows. Based on this, the concern has been raised that the skin-stripping method is not a true measurement of the drug concentration in the skin; instead, it is a measurement of the drug being deposited in the furrows instead of the stratum corneum. However, it has previously been proposed that this issue is not a concern if the skin-stripping method was properly conducted and validated. In most of the proposed methods, the remaining product that is still on the skin is carefully cleaned off, and the first few tape strip samples are discarded and not analyzed to avoid artificially high values due to the residual product on the surface. Finally, is the concern of hair follicles and follicular penetration in the stratum corneum. The skin surface is covered in varying numbers of hair follicles and sweat glands. Both of these structures form small holes in the stratum corneum barrier layer that theoretically could allow some compounds to more readily penetrate into the skin by bypassing the barrier and being partially absorbed through these "holes". Although it has been shown that some topical dermatological drug products do exhibit drug penetration in the hair follicles, the stratum corneum should reflect the major part of drug availability. Therefore, in the comparison of two topical dermatological formulations, both should have similar physico chemical properties and thus lend to the same follicular penetration.

Due to these concerns, the U.S.-FDA launched an extensive research program to develop an appropriate guidance for dermatokinetic studies. Synopses of some of the studies previously funded by the U.S.-FDA and conducted at the Department of Medicine, University of Utah School of Medicine in collaboration with U.S.-FDA regulatory scientists are discussed below.

8.2.1 Topical Betamethasone Dipropionate

A study was conducted in order to evaluate the relationship between dose and effect using the approved innovator drug product, Diprolene AF® (betamethasone dipropionate), Cream 0.05%, Scherling Corporation (Pershing et al. 1994). This product is a high-potency corticosteroid indicated for the relief of the inflammatory and pruritic manifestations of corticosteroid-responsive dermatoses in patients 13 years and older. Four methods were used to assess the dose–response relationship of topical betamethasone on the forearm of healthy subjects: duration, concentration, film thickness, and surface area. The drug uptake was measured in the stratum corneum and the resulting pharmacodynamic response was measured using a Chroma Meter®. The Chroma Meter is an instrument that reads color changes from a surface, in this case the skin, and outputs the data as standardized color scales. Several concentrations of the topical dermatological drug product were used in the study (ranging from 0.020% to 0.063%) and the duration of application time was up to 16 h. The authors concluded that a dose–response was produced with an increasing drug concentration or the duration of the application time. Also, an increase in the film thickness or surface area of the treated area did not significantly alter betamethasone dipropionate uptake into the stratum corneum in vivo.

8.2.2 Topical Tretinoin Gel

A study was conducted in order to establish a possible correlation between clinical safety/efficacy and the dermatokinetic method in bioequivalence determination for Tretinoin Gel, 0.025% (Pershing et al. 2003). Tretinoin (a metabolite of Vitamin A) is a retinoid. It is indicated for the treatment of acne vulgaris. The study was conducted in healthy patients using three approved Tretinoin Gel products: Retin-A® (tretinoin) Gel (innovator product and RLD, Ortho Dermatologicals), Avita® (tretinoin) Gel (Mylan Bertex), and Tretinoin Gel (Spear Pharmaceuticals). According to the current Orange Book, Avita (tretinoin) Gel has a therapeutic equivalence rating of "BT" and is therefore designated as not demonstrating bioequivalence to other pharmaceutically equivalent products. The generic product for Tretinoin Gel is bioequivalent to the innovator drug product, Retin-A. Based on these studies, it was determined that the drug products with similar composition and therapeutic equivalence (innovator and generic product) meet the 90% confidence interval ranges to establish bioequivalence, whereas Avita and Retin-A were found to not demonstrate bioequivalence by the normal U.S.-FDA criteria. These results are consistent with previous clinical efficacy studies of these products using patients with acne. Therefore, the authors concluded that this data did support the use of dermatokinetic studies for the assessment of bioequivalence between different manufacturers of the product.

8.2.3 TOPICAL TRIAMCINOLONE ACETONIDE CREAM

In vitro release tests, skin stripping, and skin blanching response studies were conducted in order to assess dose responsiveness and assess bioequivalence for Triamcinolone Acetonide Cream products (Pershing et al. 2002). Triamcinolone is indicated for the relief of the inflammatory and pruritic manifestations of corticosteroid sensitive dermatoses. The study was conducted in normal healthy subjects using two approved products for Triamcinolone Acetonide Cream: Kenalog®, 0.025%*, 0.1%*, and 0.5% (Bristol-Myers Squibb) and Triamcinolone Acetonide Cream, USP 0.1% and 0.5%* (Fougera). According to the current U.S.-FDA Electronic Orange Book, both products have therapeutic equivalence ratings of "AT" for topical products. Therefore, they are considered therapeutically equivalent. From these studies, the data demonstrated that with an increase in the concentration of the triamcinolone acetonide cream there was an increase in the rate and extent of drug release in vitro as well as the extent of drug uptake and skin blanching response in human skin. The authors stated that they did not find a statistically significant ($p < 0.05\%$) difference between the two sources of the 0.1% and 0.5% of the creams based on using the skin stripping and skin blanching response methods. Based on the studies, the authors further determined that the data supported the use of the dermatokinetic studies for assessment of bioavailability and bioequivalence for topical dermatological drug products.

Prior to and during the performance of these research studies, the U.S.-FDA incorporated dermatokinetic studies in two guidances for topical dermatological drug products.

8.2.4 U.S.-FDA EXPERIENCE WITH THE DERMATOKINETIC METHOD

In July 1992, the U.S.-FDA issued the Interim Guidance, Topical Corticosteroids: In Vivo Bioequivalence and In Vitro Release Methods, which included the skin-stripping method along with pharmacodynamic studies and in vitro release tests. However, it was later concluded that the U.S.-FDA had insufficient data to recommend the skin-stripping methods to document bioequivalence for topical corticosteroids. Therefore, it was removed from the guidance and the U.S.-FDA guidance was issued in the final form in 1995 without the recommendations for dermatokinetic studies.

---

* The strengths of the test products are currently designated as RLD products (Electronic Orange Book, 2009).

Three years later, in June 1998, the U.S.-FDA issued the Draft Guidance for Industry: Topical Dermatological Drug Product NDAs and ANDAs—In Vivo Bioavailability, Bioequivalence, In Vitro Release, and Associated Studies. In this draft guidance, it was proposed that the dermatokinetic method was comparable to pharmacokinetic methods used in systemically available drug products by being able to determine drug concentration in a meaningful tissue with respect to time. It was also asserted to provide information on drug uptake, apparent steady-state levels, and drug elimination from the stratum corneum based on a stratum corneum concentration–time curve (plotted as the amount/surface area against time), (Shah et al. 1998). Some of the major issues such as the use of the stratum corneum and follicular penetration were discussed in the guidance as well. In this draft guidance, pilot and pivotal bioequivalence studies were proposed, along with recommendations for performance and validation of the skin stripping technique. The metrics of the pivotal bioequivalence study were time to reach the maximum concentration (Tmax), the maximum concentration (Cmax), and the area under the concentration–time curve (AUC). In order for the test drug product to be deemed bioequivalent to the RLD, it would have to meet the 90% confidence intervals for the ratio of means (population generic means based on log-transformed data) of both the test and RLD products. The bioequivalence criteria were specified as 80.00%–125.00% for AUC and 70.00%–143.00% for Cmax.

After years of research, along with public comments and public Advisory Committee meetings in 1998, 2000, and 2001, the information and comments provided to the U.S.-FDA still raised substantial scientific concerns regarding the dermatokinetic method recommended in the draft guidance (1998) documenting bioavailability and/or bioequivalence of topical dermatological drug products. There was a high level of doubt regarding (1) the adequacy of the dermatokinetic method to assess the bioequivalence of topical dermatological drug products because the products are used to treat a variety of diseases in different parts of the skin, not just the stratum corneum, and (2) the reproducibility of the dermatokinetic method between laboratories. These issues further illustrated the complexity in developing a dermatokinetic method to determine bioavailability and/or bioequivalence for topical dermatological products. Therefore, in May 2002, the draft guidance was withdrawn (Federal Register 2002).

A considerable amount of time and resources were devoted to developing a dermatokinetic method that would be universal for all topical dermatological products, which would be comparable to but more efficient and less costly than the currently recommended clinical studies, pharmacodynamic studies, and in vitro tests. The U.S.-FDA is continually making an effort to explore the development of new methods and improvements in the current methods for documenting the bioavailability and/or the bioequivalence of topical dermatological drug products. As of 2007, the U.S.-FDA implemented its Critical Path Initiatives for Generic Products, which included potential method development for topical dermatological products. The purpose of this initiative is to document the issues and identify potential collaborative solutions with different governmental agencies, such the National Institute of Health, along with the pharmaceutical industry and academic scientists. The U.S.-FDA is aware that there is a strong need for new in vivo tools that can differentiate between the changes in the formulation that will affect the local delivery of topical dermatological products. Some of the methods being explored are pharmacokinetic studies, dermatokinetic studies (skin stripping), microdialysis, and new infrared spectroscopy.

8.3　CONCLUSION

The current U.S.-FDA regulations, guidances, and policies for submission and subsequent approval of topical dermatological drug products have led to the approval of safe and effective drug products that are currently available to the American public. Despite the U.S.-FDA's current recommendations for topical dermatological drug products, there is still an effort to evolve with the changes in the pharmaceutical industry for the development of drug products. Therefore, the U.S.-FDA is continually making efforts to explore the development of new scientific and/or clinical methods, along with improvement on the current methods for documenting the bioavailability and/or the bioequivalence of topical dermatological drug products.

REFERENCES

Critical Path Opportunities for Generic Drugs. 2007. Bioequivalence of topical dermatological products. U.S. Department of Health and Human Services. Food and Drug Administration. Rockville, MD. http://www.fda.gov/ScienceResearch/SpecialTopics/CriticalPathInitiative/CriticalPathOpportunitiesReports/ (accessed June 2009).

Electronic Orange Book. 2009. Approved drug products with therapeutic equivalence evaluations. U.S. Department of Health and Human Services. Food and Drug Administration. Rockville, MD. http://www.accessdata.fda.gov/scripts/cder/ob/default.cfm (accessed June 2009).

Federal Register. 2002. Draft guidance for industry on topical dermatological drug product NDAs and ANDAs—In vivo bioavailability, bioequivalence, in vitro release and associated studies; withdrawal [Docket No. 98D-0388]. U.S. Department of Health and Human Services. Food and Drug Administration. Rockville, MD. http://www.accessdata.fda.gov/scripts/oc/ohrms/index.cfm (accessed June 2009).

Guidance for Industry. 1995. Topical dermatologic corticosteroids: in vivo bioequivalence. U.S. Department of Health and Human Services. Food and Drug Administration. Rockville, MD. http://www.fda.gov/cder/guidance

Guidance for Industry. 1997. Nonsterile semisolid dosage forms: Scale-up and post-approval changes: Chemistry, manufacturing, and controls; in vitro release testing and in vivo bioequivalence documentation. U.S. Department of Health and Human Services. Food and Drug Administration. Rockville, MD. http://www.fda.gov/cder/guidance

Interim Guidance for Industry. 1998. Topical dermatological drug product NDAs and ANDAs—in vivo bioavailability, bioequivalence, in vitro release, and associated studies. U.S. Department of Health and Human Services. Food and Drug Administration. Rockville, MD. http://www.fda.gov/ohrms/dockets/ac/00/backgrd/3661b1c.pdf (accessed June 2009).

Pershing, L. K., Bakhtian, S., Poncelet, C. E., Corlett, J. L., and V. P. Shah. 2002. Comparison of skin stripping, in vitro release, and skin blanching response methods to measure dose response and similarity of triamcinolone acetonide cream strengths from two manufactured sources. J. Pharm. Sci. 91: 1312–1323.

Pershing, L. K., Lambert, L., Wright, E. D., Shah, V. P., and R. L. Williams. 1994. Topical 0.05% betamethasone dipropionate. Arch. Dermatol. 130: 740–747.

Pershing, L. K., Nelson, J. L., Corlett, J. L., Shrivastava, S. P., Hare, D. B., and V. P. Shah. 2003. Assessment of dermatokinetic approach in the bioequivalence determination of topical tretinoin gel products. *J. Am. Acad. Dermatol.* 48: 740–751.

Shah, V. P. 2001. Progress in methodologies for evaluating bioequivalence of topical formulations. *Am. J. Clin. Dermatol.* 2: 275–280.

Shah, V. P., Flynn, G. L., Yacobi, A., Maibach H. I., Bon, C., Fleischer, N. M., Franz, T. J. et al. 1998. *AAPS/FDA Workshop Report*. Bioequivalence of topical dermatological dosage forms—Methods of evaluation of bioequivalence. *Pharm. Res.* 15: 167–171.

Singh, G. J. P., Adams, W. P., Lesko, L. J., Shah, V. P., Molzon, J. A., Williams, R L., and L. K. Pershing. 1999. Development of in vivo bioequivalence methodology for dermatologic corticosteroids based on pharmacodynamic modeling. *Clin. Pharmacol. Ther.* 66: 346–357.

Stoughton, R. B. 1992. Vasoconstrictor assay: Specific application. In *Topical Corticosteroids*, eds. H. I. Maibach. and C. Surber, 42–53. Basel Switzerland, Kager.

Stoughton, R. B. 1987. Are generic formulations equivalent to trade name topical glucocoticoids? *Arch. Dermatol.* 123: 1312–1314.

U.S. Code. 1984. The Waxman-Hatch Act. The Drug Price Competition and Patent Term Restoration Act of 1984 (popularly known as the Waxman-Hatch or Hatch-Waxman Act) is codified at 21 U.S.C. 355 of the Food Drug and Cosmetic Act and 35 U.S.C. 271(e). and 35 U.S.C. 156 of the Patent Act.

U.S. Code of Federal Regulations. 21 Parts 300 to 499 Foods and Drugs. Revised as of April 1, 2008. Office of the Federal Register National Archives and Records Administration. Washington, DC. http://www.access.gpo.gov/cgi-bin/cfrassemble.cgi?title=200821

Index

Printed and bound by CPI Group (UK) Ltd, Croydon, CR0 4YY

18/10/2024

01776208-0005